IAN LOADER, RICHARD SPARKS,
EVI GIRLING AND BEN BRADFORD

ECOLOGIES OF SECURITY

Everyday Disorder in a Climate-Changed World

BRISTOL
UNIVERSITY
PRESS

First published in Great Britain in 2026 by

Bristol University Press
University of Bristol
1–9 Old Park Hill
Bristol
BS2 8BB
UK
t: +44 (0)117 374 6645
e: bup-info@bristol.ac.uk

Details of international sales and distribution partners are available at bristoluniversitypress.co.uk

DOI: 10.51952/9781529261226

British Library Cataloguing in Publication Data
A catalogue record for this book is available from the British Library

ISBN 978-1-5292-6120-2 paperback
ISBN 978-1-5292-6121-9 ePub
ISBN 978-1-5292-6122-6 OA PDF

Cover design: Blu inc
Front cover image: Shutterstock_netsign33
Bristol University Press uses environmentally responsible print partners.
Printed and bound in Great Britain by CPI Group (UK) Ltd,
Croydon, CR0 4YY

Bristol University Press' authorised representative in the European Union is: Easy Access System Europe,
Mustamäe tee 50, 10621 Tallinn, Estonia,
Email: gpsr.requests@easproject.com

FSC
www.fsc.org
MIX
Paper | Supporting responsible forestry
FSC® C013604

Contents

About the Authors

Ian Loader is Professor of Criminology at the University of Oxford.

Richard Sparks is Emeritus Professor of Criminology at the University of Edinburgh.

Evi Girling is Research Associate at the University of Oxford and formerly Senior Lecturer in Criminology at Keele University.

Ben Bradford is Professor of Global City Policing at the University College London.

Acknowledgements

This work was supported by the UK Economic and Social Research Council, grant number ES/S010734/1. The original study to which we refer in the book – 'The symbolic construction of crime in middle England' – was conducted under the ESRC's 'Crime and Social Order' Programme (L210252032), from 1994 to 1996. We are grateful for their support.

We have accrued many debts in the years since we have been 'back in Macclesfield'. We owe a huge debt of gratitude to all those residents of the town who let us into their lives and spoke with us, often at length, sometimes on many occasions, about the place in which they lived and for which they cared. Thanks also to all the staff and volunteers of the various statutory, third sector and commercial organizations who agreed to talk with us about their work or permitted us to observe them. We cannot name any of these people without sacrificing the anonymity they were promised. But we remain very grateful to each of them.

Warm thanks to Sergen Bahceci, Ryan Casey and Gosia Polanska for their significant contributions to the fieldwork and initial data analysis.

Finally, a big thank you to all those who have read versions of this book over the course of its preparation and supplied us with helpful suggestions: Naomi Creutzfeldt, Insa Lee Koch, Tim Newburn, Clifford Shearing, three Bristol University Press reviewers, and the reviewers at the *British Journal of Criminology*, *American Journal of Cultural Sociology* and the other journals where we have previously published work from this project. We are grateful to colleagues who have participated alongside us in panels at successive European Society of Criminology and British Society of Criminology conferences and to all who took part in the workshop on 'Ecologies of Security' that we convened in Oxford in April 2023.

Introduction: Local Security, Planetary Futures

How are we to think about everyday security in a time of intensifying awareness of environmental risks and climate disturbance? What kinds of threats and harms does ecological breakdown bring to attention in particular localities? What social meanings are attached to these troubles; to the 'old' disorders that continue to afflict daily life, and to 'new' harms that unsettle people's hopes for a liveable future? What kinds of activity do these local troubles provoke? What demands for action are made of responsible authorities now and are some such demands chronically beyond their capacity to meet? What authorities are today deemed responsible for securing the maintenance, or creation, of habitable socio-natural environments?

This book addresses these questions. We make a case for thinking about and acting upon questions of security ecologically and, in so doing, for revising understandings of the meanings and effects of disorder in everyday life. This means exploring the value of ecology as a lens through which to theorize everyday security. And it entails setting out its implications for future social investigation and political action. We set out to accomplish these tasks in the chapters that follow.

Our answers to these questions, and reasons for reappraising how we think about security today, are inspired by and grounded in four years of research in one particular place. That place is Macclesfield, in north-west England. Macclesfield is a town of 53,000 people. It is about 30 kilometres south of the nearest large city, Manchester, and sits between the Cheshire plain to the west and the Peak District hills to the east. Macclesfield has much in common with other towns of similar size in the UK. It is a place of moderate prosperity, though also one with entrenched social difficulties and socio-spatial inequities. It is in a national context relatively safe – and mostly understood as such by its residents. But it is not without problems of crime, disorder, insecurity and vulnerability, and attendant demands for protection. The population is a mix of residents who have lived in Macclesfield for generations, alongside those who moved there to work for some of the town's big employers (such as AstraZeneca) or who have chosen to make Macclesfield their home and commute to nearby conurbations. There are a small number of established and recent migrants (some of whom have found refuge from wars and conflicts far away), but demographically the town remains predominantly white British.

Macclesfield is, at first blush, an odd place to situate an argument about the changing meanings of (dis)order and (in)security in a time of climate crisis (Read, 2022; Otto, 2025). It is in global terms prosperous and secure. The town as a whole is not at risk of intense flooding, though certain streets might become so, nor is it directly vulnerable to rising sea levels. It is not at present exposed to extended drought or extreme fires, still less the loss of arable land and desertification. It is, conversely, not entirely immune from the effects of a warming planet upon weather patterns and local ecosystems, nor from future climate-related social conflicts: nowhere is.

Nor did we find its residents to be especially environmentally conscious, though the town has many active environmental groups concerned to live and act sustainably and protect local habitats. For many of those citizens and their wider networks, Macclesfield is and should be a place where concern for the environment is a prominent feature of everyday and civic life (Steele et al, 2021; cf Norgaard, 2011). Many of the problems that concern agencies and residents of the town also seem at some remove from ecological questions. During our fieldwork the police and local authorities were engaged by problems of vulnerability and exploitation, and risks of violence in domestic settings. Residents were most commonly victimized by online fraud and report being worried about drugs, and neglected youths, and enduring problems of burglary and car crime. These seem like a mix of 'old' security questions, prompting familiar calls for police attention, and some new ones with a specific 'cyber' twist, a long way from the environmental harms that preoccupy green criminologists and climate activists. So, in what senses are these also ecological questions?

That was then ...

We have visited – and studied – this town before. Thirty years ago, in the mid-1990s, three of the present authors (each then based at Keele University) investigated public concerns and talk about crime in Macclesfield and its relation to people's sense of place. That study resulted in a book-length account of the connections between 'crime-' and 'place-talk', called *Crime and social change in Middle England* (Girling et al, 2000). The study took place in a context where anxiety about crime had come to be a normal part of everyday routines and consciousness and a prominent and highly charged topic of media discourse and political contestation (Garland, 2001; Simon, 2007). 'Fear of crime' was a salient feature of lived experience, as well as a shared object of governmental intervention and criminological concern. In the Ipsos-Mori Issues Index for October 1994 (the month our fieldwork commenced) 41 per cent of those questioned listed 'crime' as one of the three most important issues facing Britain. We know now (but could not know then) that the mid-1990s was the moment of peak anxiety about 'crime', at least as measured by the British Crime Survey (Tiratelli, 2025). It was also at the peak of actual crime levels, as measured by the same source (Office for National Statistics, 2025).

We chose Macclesfield because it was a moderately affluent and relatively safe town; in sharp contrast to the marginalized, high-crime neighbourhoods of big cities that have long been the staple of criminological research. It was never our intention to depict the town as a place of tranquillity, untroubled by many of the same problems as other places and immune from the greater upheavals of the late modern world. Rather, we wanted to understand how questions and controversies about crime, policing, social order, fear, and the governance of and accountability for those problems, were encountered and experienced *in situ*, in the fairly un-dramatic context of one medium-sized town.

We spent two years, using a range of qualitative methods, trying to find out how experiences of, and talk about, crime shaped local social relations. Much of what people told us was not in itself hugely surprising. They had concerns that were fairly characteristic of the period, and not dissimilar from those that regularly appeared in media discourse and political talk about crime in Britain in the 1990s. Their worries focused on burglary and theft of, or damage to, vehicles, which they suspected was often the work of criminal outsiders. We found adult residents agitated about young people 'hanging around' in public space, causing disorder, or consuming drugs. We encountered a structure of local feeling preoccupied with the felt disappearance of the police from community life and angrily (perhaps also nostalgically) demanding visible police protection. We found people coming to terms – for the most part positively – with the arrival and rapid proliferation of CCTV.

These worries were shaped by people's sense of place – their situatedness in the economic and social hierarchies of Britain in the 1990s, and the location in which they resided, and its past, present and anticipated futures. We documented and analysed the ways in which crime worked as a condensing symbol of wider social and cultural change, both national and global, from which local life was felt to offer some degree of insulation. Macclesfield was valued in part because it provided a protected space. Crime was troubling in part because it materially and symbolically threatened that space. Macclesfield in the 1990s felt 'defended' – a place of relative, if fragile, safety and prosperity whose residents sought protection against unwanted incursion from outside – whether in the form of travelling criminals or dismaying moral turbulence and cultural trends.

We hoped that our interpretation of what people told us created some new ways of seeing – focused largely on how the experience of crime, disorder and locality was being reshaped in conditions where people, capital and culture were circulating with increasing speed and intensity across the boundaries of place and nation. In this respect, that original study took place in two structuring contexts: first, we very consciously situated it against the backdrop of what was increasingly assumed then by social theorists and politicians alike to be the inescapable fact of globalization (Massey, 1991; Blair, 2005). But second, as has since become apparent, our first study took place during a period of sustained economic prosperity that economists started referring to as 'the great moderation' (Benati, 2007). That period was not to last.

... and this is now, or at least lately

In 2019, we returned to Macclesfield. We did so in the knowledge that much had changed in the intervening quarter of a century. This was so in respect of marked technological, social and political transformation: the arrival of the internet and social media, the 2008 financial crash and austerity politics, Brexit and its aftermath, intensified concern about global warming and, shortly after our fieldwork started, the COVID-19 pandemic. But much had also changed in terms of levels and patterns of crime, and modes of security governance. Two full decades of declining rates of burglary, car theft and numerous other volume offences in many affluent countries have served to turn down the heat generated by the preoccupations of our original study. According to one wave of the Ipsos-Mori Issues Index conducted in the midst of our fieldwork in 2021, the biggest issue for the British public was climate change.[1] Crime, though of fluctuating and sometimes acute interest throughout the preceding two decades, was at this point not in the top ten of people's concerns.[2]

Against that backdrop, we made a 'punctuated revisit' (Burawoy, 2003: 669) and set out, once again, to investigate and understand the place and meanings of in/security in the everyday lives of the people living or working in this unremarkable English town. We were interested in the crime and security concerns of differently situated groups of people, and the demands these prompted from authorities. What crimes loom large in everyday experience and consciousness? Was crime any longer the pre-eminent 'security' concern troubling local people? If not, what other fears and anxieties now shape everyday life? We also explored again the extent and manner in which these threats and demands are mediated through people's sense of place and attendant feelings of belonging. How are local experiences of security, and people's sense of what action can or should be taken to foster and sustain civil association, tied up with anxieties and hopes for the future of the places they inhabit, whether at a local, national or planetary scale?

To address these questions, we deployed a range of methods within a study whose overarching sensibility remained ethnographic. Each of these methods provided 'exposure' (Small and Calarco, 2022) to different aspects of local people's experiences of, and demands for, security and the practices of security governance of various state and civil society actors. We carried out two community surveys. Both used face-to-face interviews undertaken by a social research company. The first survey (n = 416) was conducted between May and August 2021 as pandemic restrictions eased; the second (n = 502) took place in September and October 2022; different participants were interviewed for each wave. Many questions were common across both surveys, though new questions were added to the second in response to the themes emerging from the interview and observational research. About 2 per cent of the town's population took part in the survey, with the sample spread across age, gender and social class.

We also held in-depth conversations with residents individually and in small groups – in their homes, in public rooms, sometimes while walking or cycling with them. These included 41 individual interviews, 15 focus groups and seven biographical interviews ranging in length from 45 minutes to two-and-a-half hours. Participants were recruited by a combination of responses to a leaflet delivered to every residential address in the town and snowballing from contacts made during our access negotiations and observational work. We aimed to speak to people from as many socio-economic groups as possible, though in our final sample residents over the age of 30 were over-represented. We carried out interviews, informal discussions and observations with actors from local government, private security providers and the voluntary sector. We interviewed 23 formal and informal security providers including elected officials, key actors in the local voluntary sector, town centre security managers and police. We spent over 300 hours following officers and Police and Community Safety Officers in their everyday work around the town (on day, night and weekend shifts) as well as observing civil enforcement officers.

We also carried out formal and informal observation of public spaces. Members of the research team engaged with various local groups – including youth charities, foodbank volunteers, litter-pickers and community gardeners – and we attended a wide range of local events. We observed meetings of local authorities (Cheshire East County Council and Macclesfield Town Council) both online and in person. We analysed how the relation between security and place is articulated, debated and visually represented in social media. On Twitter (now X), we collected and analysed police tweets from 13 Macclesfield-relevant police accounts (n = 1,389) dating from April 2018 to July 2021. We also analysed tweets about Macclesfield and its places (n = 7,660) from November 2019 to April 2022. On Facebook, public town/neighbourhood-related pages were monitored throughout the period of the research, with more intensive analysis in spring 2020, spring 2022, and early July 2022 until the end of January 2023.

We analysed official documents pertaining to everyday security – local authority consultations and strategy/policy documents and proclamations of the police and elected Police and Crime Commissioner, for example – as well as tracking local planning applications and the objections they prompted. We also elicited a small number of audio/photographic diaries from residents – largely in response to our inability to conduct face-to-face fieldwork during the pandemic – and took hundreds of photographs to aid our understanding of the relation of everyday security to people's sense of place. At the conclusion of the research, we held an all-day deliberative workshop attended by around 60 residents, as well as local councillors and representatives from Cheshire Police and the Office of the Cheshire Police and Crime Commissioner.

Embarking on the revisit, we were much more uncertain – and curious – about what we would discover when, across different settings and using this range of methods, we posed the question: what troubles people living round here nowadays? We were also conscious that in the intervening 25 years we had variously changed – in experience, age, seniority, outlook, theoretical orientation – and remained alive to how this affected our enquiries and interpretations (Burawoy, 2003; O'Reilly, 2012). It no longer seemed to us that we could confidently anticipate what would fill people's inventory of concerns, and how they would make sense today of the relation between their local experiences of everyday in/security and the fault lines of the wider world.

What we in fact discovered were some marked shifts in the local meanings of disorder that seem to have developed since our research in the mid-1990s – shifts that we initially discounted, or squeezed though the theoretical frame of the first study, but slowly came to understand as changes marked and felt among residents of this town, not imposed upon them by us. People still sometimes talked about burglary and car crime – though volume property crime has for most residents become an episodic experience, less constitutive of their sense of place. There was less public discourse about groups of young people in public space, though such presence was now often coded as more serious – a sign of neglect and vulnerability, rather than 'kids being kids'. Residents still worried a lot about drugs – in terms of what drugs ominously intimated about veiled suffering and exploitation in the town, and the impact of drug detritus on the quality of local neighbourhoods or parks. The terrain of troubles and agitations remained, in these respects, familiar and seemingly well-understood within the frames of socio-spatial and environmental criminology.

But in their discussions with us, and one another, residents frequently focused, perhaps with a new intensity, on a range of chronic harms to local environments. People spoke about graffiti, vandalism, potholes, litter, abandoned buildings and unkempt public places – streets, parks, whole neighbourhoods. They reported to us a sense of a town centre in decline and feared for its future. Cars emerged as significant agents of everyday disorder, in terms of congestion, speeding, parking, air pollution and daily conflict on the school run. People worried about the environmental impact of growth and development while also expressing concerns about the affordability and availability of housing for young people. Some actively sought to protect local ecosystems from danger or elimination. Residents of this town still wanted police protection, and for policing to be visible. But much local anger and puzzlement was now directed at what people regarded as an absence of guardianship on the part of the local authority, and thus failures in basic urban governance. They worried that no one beyond the town seemed to care about, or to be looking after, it. These worries could generate new kinds of censure and blame, new folk devils if you like. But they were also attended less by discourses of 'us' against 'others' than by sentiments of complicity; the uneasy feeling that – in respect of many current concerns about everyday security – people (like us) have become agents of our own peril. These harms matter – we slowly came to realize – because they are entangled with people's sense of the quality and sustainability of places they care about and inhabit. They pushed our thinking in an ecological direction that we had not originally anticipated (Loader et al, 2023).

But if people's concerns about disorder had changed, so too had the structuring contexts that condition everyday life in this corner of England. The second study unfolded in a social and political context characterized by 'an ambient kind of austerity' (Harris, 2022) – over a decade of straightened public investment the indicators of which are observed by our participants in the degraded and neglected fabric of the public environment (see also Hitchen, 2016). This is the backdrop against which people are immersed in their (local) environments: they sense their street, neighbourhood or town in ways that are attuned to certain chronic signs of disrepair that impede their capacity to go on comfortably within it or feel confident about its future prospects (see, further, Erker et al, 2023).

But our investigation also – in all kinds of obvious and not so immediately obvious ways – took place in the shadow of climate breakdown, or what Latour (2017: 8) calls profound ecological mutation. The background focus of our first study had been 'the global' (Girling et al, 2000: ch 1). Our concern was with how globalization permeated and shaped local social relations, and people's sense of place and its prospects, in this one English town. That problematic has by no means gone away. But globalization is an anthropocentric process. It is 'chiefly concerned with the existence of markets for commodities and circuits of communication that place *human life* and human-centred institutions at its heart' (Matthews, 2023: 147; emphasis in original). Second time round, our enquiries also occurred against the more fundamental background condition of 'the planetary' (Chakraborty, 2021). This is the implicit sense – detectable in daily practices and sensibilities, not just in local environmental activism – that the security and liveability of a place is bound up with the fragile condition and future fate of a wider socio-natural environment, of which we are each a part, and upon which we each depend – 'a deeper appreciation of the materiality of the Earth and ourselves as human' (Mould, 2023: 4).

These, we suggest, are the principal contexts against which some previously unremarkable objects and practices are identified as features of the local 'harmscape' (Berg and Shearing, 2018) and problematized in novel ways.

These concerns, however, vie today with a very different, and altogether noisier and more visceral, politics of place. Trust in democratic political authority stands at a perilously low ebb (Curtice et al, 2025). Authoritarian politics appears to be globally and nationally in the ascendant, and autocrats contend seriously for, or occupy, political office. Concerns about asylum-seekers and international migration dominate public discourse, often in ways entangled with anxieties about crime and stern demands that the borders of 'our' community and nation are defended. In the October 2025 wave of the Ipsos-Mori Issues Index, 'immigration/immigrants' was the top public concern, with 47 per cent of mentions (though 'crime/law and order' was joint ninth with 9 per cent). These preoccupations are stoked in online silos (notably, Elon Musk's X) and by the agitations of Reform UK and other far-right activists. Racism and xenophobia have returned as regular and 'acceptable' features of public discourse. Conversely, the apparent consensus on the need to tackle global warming and meet 'net zero' targets has come under increasing strain. Climate mitigation is attacked from the political right as a hoax or diversion from more pressing issues and seems to have dropped down the list of public concerns: 'pollution/ environment/climate change' shared ninth place in the October 2025 index.[3] In his reporting from Blyth in north-east England – an economically stressed town in uncertain transition, made up of 97 per cent British-born residents – James Meek captures that ambient mood as follows:

> The town, or at least a significant number of people who live there, has become obsessed with immigration – immigration as a source of disorder, discomfort and disruption; an abstract, visceral sense that outsiders make unfair claims on places, benefits and opportunities to which they have no birthright, and which should therefore be denied them; a belief that the removal of strangers whose appearance and manner mark them out as different would cause no new problems and make the natives happier. (Meek, 2025: 21)

For the most part, these hostile concerns did not feature prominently in our research, for reasons that have only partly to do with Macclesfield's 'whiteness' and the relative absence of foreign migration to the town. This would also seem to have much to do with how rapidly the immigration issue has intensified in public agendas over the last few years. If we were to return to the field (yet) again now, it would doubtless figure more prominently, albeit (we suspect) quite variably. What we encountered, instead, is a largely alternate story about the condition of England, a contending account of the mundane troubles and anxieties that afflict people nowadays, and their hopes for what might be done to address them.[4] By telling that story, our task is to 'expand the bounds of what is imaginable … to build a common affect, a desire for certain ideas, values and practices to exist precisely among those who will be their constituencies in the future. What is necessary is not to just witness but also participate in this shift' (Bhan et al, 2025: 131).

New ecologies of security

This book is *not* an overview or summary of this research – though we elaborate and reflect on its themes in the pages that follow. The detailed findings and analyses of our multi-method data can be found elsewhere.[5] Nor does it offer a fine-grained account of how differently situated people living in the town experience safety and insecurity in all their dimensions nowadays. What follows, instead, is an extended essay on how – and why – this study has led us to seek to bring the sub-fields of environmental and green criminology into closer conversation with a view to crafting a new ecological orientation towards questions of security and how to act upon them.

In this respect, the book is an exercise in 'analytical re-description' of the themes to have emerged from our research. In *New Urban Worlds*, Simone and Pieterse (2017) suggest that such an approach 'entails re-describing existent conditions as components of a process that might be taking place *right now*, but which is opaque, occluded or rendered inoperative'. 'It is rendered inoperative', they continue, 'simply because we are seeing and engaging the realities examined in a particular way'. 'So there is a *doubleness* here', Simone and Pieterse argue, 'that involves a sense of aspiration, of making things different, but also of seeing in what exists something other than what we think we are seeing' (2017: 11; emphasis in original).

What might this mean in the present case? Our point is not that all questions of everyday security are consciously entangled with, or can somehow be reduced to, the threat of ecological collapse. That would be a misleading claim that did violence to the scope and variety of people's concerns and their sources. Much of what troubles Macclesfield residents on a daily basis exists at some remove from the climate crisis, which bears little if at all on how they make sense of the impact of crime and disorder on the town and their daily lives. But it is, at the same time, possible to discern a distinct 'green' inflection to people's ways of voicing their concerns and imagining remedies to local troubles. On this view, problems such as street-drinking, litter, neglected buildings, the condition of parks or fate of the town centre, badly parked cars, pollution, and so on matter because they *are* caught up with people's sense of the quality and sustainability of the places they inhabit, and with urgent questions to do with who will act, and how each of us should act, in the face of threats that erode the boundary between the very local and the planetary.

Our task in this book is to 'centre' this interpretive framing and bring to the fore the ways of seeing and acting that it discloses. Our aim is to render explicit the implicit sense – discernible in the routines and outlooks of daily life, not just in local environmental activism – that the security and liveability of this place is bound up with the fragile condition and future fate of a wider socio-natural environment, upon which we each depend. The book is, in these respects, an attempt to re-describe the present in order to 'compose urban knowledge of *what can be as well as of what is*' (Simone and Pieterse, 2017: 11; emphasis in original).

Across our two studies, Macclesfield has never represented one of the 'privileged material research objects' on which urban criminology typically draws (Krause, 2021). It is not a multiply deprived, high-crime, inner-city neighbourhood. That has remained part of its appeal. During the 1990s, the town was a surprising, but for that reason revealing, location for understanding social anxieties about crime. Three decades on, it is a surprising, but for that reason revealing, site for grounding an argument about the meanings and effects of everyday disorder in circumstances of durable austerity and ecological breakdown. In both cases, this one medium-sized English town has served as 'an important site for thinking about how the bigger stories of change and disruption are experienced, lived, negotiated and responded to' (Steele et al, 2021: 21). Or so we hope to show in these pages.

We develop the argument as follows. In Chapter One, we offer a brief history of the concept of ecology in the human sciences, teasing out its origins and subsequent uses in the social analysis of urban crime and disorder. In Chapter Two, we outline the elements of a new ecology of security, paying attention to both of those terms. To do this, we bring the concerns of environmental and green criminology into dialogue with the aim of identifying connections between the local disorders and concrete moral engagements of everyday life and some larger questions pertaining to the state and prospects of living in a climate-changed world. The chapter serves as a 'hinge' to the more situated analysis developed in the remainder of the book. In Chapters Three to Five, we describe the concerns reported in our research with respect to the degraded, littered state of local socio-natural harmscapes, the decay of physical and social infrastructures, and the damaging impact of cars on the town's social ecology. In each case, we tease out the meanings and implications of seeing these as ecological problems. We begin Chapter Six by outlining some forms of mundane action that residents take in respect of the matters of concern around which they assemble and consider their value as practices of environmental care and repair. We then identify the kinds of responsibility that are demanded of local authorities (and the urgent complaints about failing infrastructures of care) and consider what follows if we re-conceptualize these practices and concerns under the umbrella of the ecological governance of place. A brief conclusion brings together the principal themes and implications of our analysis.

ONE

A Brief History of Ecology (and Crime)

We set out, in 2019, to study the ways in which people's concerns about the security of their everyday lives, and their worries and experiences about crime, disorder and related troubles, arose in one English town. We did this not for the first time, but with some new dimensions, attuned, we hope, to the shapes that those sensibilities and feelings had come to assume these days (Loader et al, 2023). From the outset of the first project, now some 30 years ago, we wanted to explore how people's concerns on these matters were involved with their sense of place – the town as a place, their place within it, and its place in a turbulent and uncertain world. We have never doubted that people's connections with place were in some important respects particular, tied to local landscapes, buildings, history, culture and relationships; but equally we have taken it that these are not so unique as to have no bearing on the experiences of other people in other places. In our more recent study we have noticed, pretty much for the first time, that many of the concerns that people voice have an environmental aspect, and that this informs their sense of what they, or others, or someone, might do about it all.

For these reasons we have rather grandly introduced the term 'new ecologies of security' to signal the ambitions of this book. We do this quite consciously, albeit in the knowledge that our use of these words stands in a somewhat quizzical and oblique relationship to some of the traditions on which we draw. Place, environment, community and relationship are all ideas that loom large in the history of the uses of the idea of 'ecology' in both its (some would say proper) natural scientific and its social and political applications. Among our reasons for foregrounding this idea now is that coming to terms with the contemporary 'green' inflection of people's expressed concerns for the present and future well-being of their habitat requires us to bring certain bodies of thought together in new ways.

Despite their seeming verbal connections, 'environmental' criminology (work that primarily has to do with crime, risk, fear and place) and 'green' criminology (work that primarily has to do with crime, harm and the natural environment) have historically developed largely separately. Their practitioners do different kinds of research, refer to different authorities, attend different conference sessions, quite often believe that 'theory' performs different (and sometimes incompatible) functions, and may have quite divergent views on the political and social tasks of criminology, too. Both, however, make important reference to the idea of ecology or 'ecological' approaches. What do they each mean by this? If we can tease this out, we may be able to interpret some of our observations more productively. People's worries and concerns about aspects of their habitat often arise in the context of discussions about troubles and disorders that have very specific relations to places, and that are often hyper-local in character, even if, as we believe, they invoke connections to much wider changes and fears for the future. We therefore need both an 'environmental' criminology that comprehends the idea of harms or threats to the natural environment among its inventory of concerns; and a 'green' criminology that understands something about mundane security in the settings of everyday life; and the starting point is that these are both 'ecological' questions.

We therefore begin by glancing back into the history of the uses of the idea of ecology in a number of fields – briefly in respect of its origins in biology and environmental studies; at a bit more length in respect of criminology (and related areas of anthropology, sociology and urban studies); and finally in relation to contemporary debates about human action and climate breakdown and environmental crises. Ecology, as we shall see, is a very complex word; one used in multiple ways, and sometimes frustratingly loosely and misleadingly. Since we cannot avoid it, however, we had better decide what to do with it.

The ecology

The basic premises of ecology as an idea and field of study are expressed in more or less identical terms in a wide range of sources. The coining of the term is usually attributed to German zoologist Ernst Haeckel in the 1860s. Ecology is a neologism that connects two Greek terms – *Oikos* (home or household) and *Logos* (reason or knowledge). The eco- in ecology thus has the same source as in eco-nomics. It involves, perhaps poignantly for our own appropriation of the term, the science or study of homes, or as we would more usually say, habitats or environments (Schwartz and Jax, 2011: 145; see also Benson, 2020).

Closely following Haeckel, standard definitions affirm that ecology is the study of 'the relationship between organisms and their environment' (Smith and Pimm, 2025). Such definitions commonly hasten to elaborate that these relationships are inordinately complex. They refer to the behaviours of individual organisms but also to communities, both within and between species, and to social as well as physical aspects of environments. At different levels of analysis, therefore, ecological inquiry might refer to the reproduction, breeding success and populations of specific species, but also to the entirety of living things present in an environment (the 'biomass'), whose interactions with one another, and with the 'abiotic' components of their surroundings, constitute an 'ecosystem'. Space, relevance and our own limited knowledge forbid us from elaborating much further.

Suffice to note that even seen from within strictly biological frames, these interactions and properties of ecosystems necessarily also comprise humans. This applies in the obvious and generic sense that humans as a species reshape, cultivate, predate upon, create and destroy the habitats and populations of other species, and thus their own, to an overwhelming extent. More specifically, the degree to which this is noticed, and the rapidity of the changes we have wrought, have accelerated to the point that, as the authors of the *Britannica* article note with wry understatement: 'the most pressing problems in human affairs – expanding populations, food scarcities, environmental pollution including global warming, extinctions of plant and animal species, and all the attendant sociological and political problems – are to a great degree ecological' (Smith and Pimm, 2025).

Yet the appropriation of the idea of ecology to apply to human worlds goes back almost as far as the term itself and long precedes its contemporary application to our current crises of environmental degradation and climate breakdown. Ever since ecology began to displace earlier discourses like natural history and taxonomy as a way of conceptualizing the natural world and its dynamic processes of continuity and change, students of human, social phenomena – especially perhaps of cities – have seen analogies between its ideas and their interests. Ecology is intrinsically a study of relationships, not of entities in isolation. It envisages a complex array of nested levels of inquiry (organism, community, population, ecosystem, biome). Ecology tends to lend itself to systemic ways of thinking – the interactions between species and environments in an ecosystem pose intricate challenges of modelling and prediction and rarely reduce to linear let alone mono-causal explanations. Some ecologically inspired social theories, perhaps less fashionable now than formerly, interpret social divisions and stressed social relations as loops of negative feedback in systemic processes (Bateson's idea of *schismogenesis* [1972], for example – one of the points of origin of the common expression 'vicious circle').

Others, more optimistically, have seen in the idea of a 'social ecology' a paradigm for social cooperation and development without the need for direction by a central power (Bookchin, 1982). However, for all the potential fascination and promise of these ideas, we must first focus on the uses of 'ecology' within the tradition of studies of crime and city life.

'Ecology', crime and the city

Studies of the spatial distribution of crime in cities, and of persistence and change in differences between sites in the city, form a central tradition in the modern history of criminology, perhaps indeed the single most salient one. One leading authority defines the 'ecological' character of such studies thus:

> Social (or human) ecology may be broadly defined as the study of the social and behavioral consequences of the interaction between human beings and their environment. It specifically explores the causes and consequences of processes of segregation – the emergence through selection of environmental differentiation along key dimensions such as population composition and land use. It investigates how exposure to different environments (area- and place-based differential social organization and activities) influences human development and action. The social ecology of crime is the study of one particular behavioral outcome of these processes, the violation of rules of conduct defined in law. (Wikström, 2009)

It is notable here how closely Wikström's formulation mirrors the standard definition of ecology in biology – the study of interactions between organisms (humans) and their environment. This is a definition that takes its ecological heritage seriously (not to say literally). Wikström goes on to offer an account of levels of explanation in ecological studies of crime that also in some degree mirrors the ladder of objects of study in ecology: 'the advancement of a fully developed ecological perspective on crime (a full understanding of the role of the human-environment interaction in crime causation) requires a better integration of environmental and individual approaches in the study of crime causation' (Wikström, 2009).

Clearly, our concerns in this book are not quite of this kind. We focus on people's accounts of everyday security and insecurity, not on causes of crime. We will return later to the issue of whether and how this too may be considered to be an 'ecological' question.

One of the key sources for the tradition in which Wikström locates his concerns is the so-called Chicago School (work conducted in or influenced by the Department of Sociology of the University of Chicago from the 1920s to 1950s, and most often about the city of Chicago itself). In many accounts, the term 'ecological' in the context of criminology is used as more or less synonymous with 'Chicago School' or as an umbrella term for area-based studies of social organization and disorganization (see, for example, Tibbets and Hemmens, 2018).

At its broadest, the 'ecological' inspiration for the tradition of studies of urban scenes in criminology is provided by the complex of relationships that people experience and encounter with the social and physical environments through which they move. Bottoms (2012: 529) expresses a preference for the term 'socio-spatial criminology', perhaps to distance contemporary research from any over-extension of the biological analogy in 'ecology', but also to encompass the great breadth of economic, cultural, demographic and other dimensions at play. In an essay whose title alluded to 'environmental' criminology (a source of potential confusion as Bottoms acknowledges [2007: 529, n 2]) Bottoms and Wiles see socio-spatial criminology as encompassing 'the study of crime, criminality and victimization, as they relate first to particular *places*, and, secondly, to the way that individuals and organizations shape their activities *spatially*, and in so doing are in turn influenced by *place-based* and *spatial* factors' (Bottoms and Wiles, 2002: 620; emphasis in original).

Historically, a key source for the association between 'ecology' and criminology is the work of Robert Park (1936; 1952). Park deploys the e-word throughout his work in various ways. Ecology is a shorthand for the city as a complex and variegated set of habitats (or ecosystems as we might be tempted to say nowadays). It presupposes complex part/whole relations between cities and their constituent communities. Park envisages denser relations within communities than between them (the so-called theory of 'natural areas', often defined in ethnic terms). Such ideas are sometimes used to suggest analogies between the spread of problems such as crime, and examples of invasive species or contagions in natural systems. The latter may sound somewhat outlandish today, or at least seem to depend on the over-application of the biological analogy in reference to 'ecology'. As Bottoms notes, subsequent contributors to the Chicagoan tradition, most notably Shaw and McKay (1942), were more circumspect about this, and more explicit about the social and economic basis of change and difference between urban areas. Nonetheless, questions that animated Park's work, such as the emergence of residential segregation, boundary maintenance, and contests over the meaning or ownership of social space, have remained important in shaping the zones and divisions that contemporary observers of cities like Chicago still encounter (Sampson, 2012).

Subsequent generations of Chicagoan sociologists and their allies have therefore developed these ideas to study and map the succession in dominance of particular groups within given territories via housing choice (Shaw and McKay, 1942), exclusion (Wilson, 1987; Massey and Denton, 1993), gentrification, 'flight' (Taub et al, 1984) and other mechanisms that document and problematize embedded inequalities. Within this tradition the deteriorating neighbourhood has long occupied a special place of concern (for example, Taylor and Covington, 1988; Skogan, 1992; Walker, 2011).

Small wonder then that Wikström places such emphasis on the place of segregation within the tradition of 'ecological' criminology: segregation cements social difference, and locks in differences in people's exposure to crime and other social problems.

Where is the environment in 'environmental' criminology?

Wikström (2009) and others (Walker, 2011) depict 'environmental criminology' as a subset of the ecological (or 'socio-spatial') tradition. Arguably, however, it is one that carries a certain tension with the more structural sociological ambitions of the wider field. For socio-spatial criminologists and urban sociologists influenced by the Chicago tradition, the key points of reference are the city and the neighbourhood or area; and their concerns include the differences in life chances and opportunity that arise between such localities, with all the political and policy challenges that they pose.

By contrast, in its most prominent articulations 'environmental' criminology tends to refer to more micro-level features of places, with particular emphasis on the physical properties of the urban scene (buildings, streets, intersections, transport hubs, licensed premises, light and dark places and spaces, and many more) as providing different levels of opportunity and risk (for example, Cozens, 2011; Ceccato et al, 2025). Where the socio-spatial tradition tends to emphasize structural social aspects of neighbourhoods as providing the deep context for action, the new environmental criminology focuses on the scope for intervention to reduce crime in the here and now by more rational deployment of police attention, street lighting, 'target hardening', defensible spaces, and other forms of opportunity-reduction. Cohen and Felson influentially affirm that '[u]nlike many criminological inquiries, we do not examine why individuals or groups are inclined criminally, but rather we take criminal inclination as given and examine the manner in which the spatio-temporal organization of social activities helps people translate their criminal inclinations into action' (Cohen and Felson, 1979: 589). Cohen and Felson nonetheless still assert that these are 'ecological' questions in that they refer to 'the fundamental human ecological character of illegal acts as *events* which occur at specific locations in *space* and *time*, involving specific persons and/or objects' (Cohen and Felson, 1979: 589; emphasis in original).

For its proponents this focus on timing, spacing and location gives the environmental turn in criminology a policy payoff that the field has rarely displayed before (Clarke, 1995). For these reasons, situational crime prevention, routine activities theory, and the broader terminology of crime prevention through environmental design have come to be regarded by many as essential tools for police managers, urban planners, architects and other professionals (Crowe, 2000; Cozens et al, 2005). For more sceptical observers there is a danger of addressing neither structure nor agency – a withdrawal from the challenge of explanation in favour of a technocratic orientation to intervention (Hope and Sparks, 2000).

In this context 'place' tends to refer to specific locations, often at a very fine-grained level of analysis, that have attributes construed as criminogenic, or that give rise to exacerbated fears of crime, and which then tend to be avoided by some and perhaps frequented by others. The point is to identify the unit of analysis that is amenable to documentation and intervention (Weisburd et al, 2009). In one influential (some might say notorious) line of descent within this family of approaches, the lack of visible guardianship over particular sites tends to initiate a pathway of disorder and decline (Wilson and Kelling, 1982). Here, fixing 'broken windows' may be seen as requiring not simply a literal new window – if it is this at all – but rather a policing strategy that foregrounds swift and effective action against begging, loitering, drug-taking or other disorders viewed as implicated in the deterioration of the place in question. In this regard, however, whether and in what respects the enhancement or protection of the environment is a core concern for environmental criminology is a contentious question.

Greening environmental criminology?

Generations of police officers, local government workers, shopping centre administrators and many others have found research in environmental criminology useful in various ways. It is not hard to see why those who are charged with pragmatic intervention in certain kinds of 'security project' (Valverde, 2011) should do so. We fully acknowledge this, and taking issue with it is not our purpose here. However, we would suggest that concern about the environment in environmental criminology in the main takes certain limited and instrumentalized forms.

Environments matter here to the extent that they engender increased risks of crime or are degraded by actions deemed disorderly or unsightly (the littering, loitering, public drinking and so on). One of the main policy sequels of 'broken windows' thinking is sometimes known as 'quality of life policing', an effort to recover a sense of public order in specific locales by cooling 'hot spots' of crime (Braga and Bond, 2008). These tactics are intensely debated in terms of their effects on many aspects of police roles and conduct, questions of equity in police relations with members of marginal and disadvantaged publics, and hence on the legitimacy of the police and other public authorities (Harcourt, 2001; Weisburd et al, 2011). Undoubtedly, acting on these premises has effects on the priority that authorities give to tidying up, cleaning up, restoring and beautifying urban scenes. In all these senses environmental criminology undoubtedly has important environmental effects. However, in what follows we want to show that the participants in our study often have something rather more far-reaching than this in mind when they speak about troubles in, or threats to, their local environment, and that they indeed envisage connections between the particular disorders and forms of damage that preoccupy criminologists and environmental futures. Can we extend the imagination of environmental criminology to deal with this? Is environmental criminology still more concerned with whether a front garden is a defensible space than with whether it is an ecologically impoverished one, forgetting the fact that the latter conception has distinct security implications?

Something similar seems to be true also for the longer tradition of studies of the 'social ecology' of crime that derives from the Chicago School. These are deeply interested in socio-spatial variations in crime and victimization and their relation to other aspects of urban life. Undoubtedly these matters can intersect with concerns about environmental questions. Sampson has recently documented different levels of exposure to lead poisoning, for example, in the context of his broader, magisterial renewal of sociological interest in differences in crime and other vectors of inequality in neighbourhoods in the city of Chicago (Sampson, 2012). Socio-spatial criminologies carry implications for urban planning and governance and have long been recognized as doing so. Yet this is some distance from integrating a developed concern with environmental issues into criminologists' inventories of concern. As we have indicated here, the 'ecological' tradition in criminology borrows certain important formal and conceptual features from ecology. It is interested in the relations between individuals and environments, and in systemic patterns in the distribution of groups and communities across socio-spatial boundaries. Features of the urban landscape – streets, plazas, rivers, parks – certainly fall within the horizon of interest of these traditions; but whether criminological ecologies have (or need have) any developed interest in the ecological dimensions and implications of those places is a moot point.

In an intriguing footnote, Bottoms indicates that the change in title of his contribution between successive editions of the *Oxford Handbook of Criminology* from 'Environmental Criminology' to one indicating a more general interest in socio-spatial questions was in part because he had come to feel that the term 'environmental' should be reserved to refer to 'green' issues (Bottoms, 2007: 529, n 1). This is a characteristically acute and prescient thought. No doubt one could dispel a good deal of confusion if it were possible to convince criminologists to stop using 'ecology' and 'environmental' in excessively broad and poorly defined ways.

Our view, however, is in some respects the reverse of this. Far from keeping 'environmental' and 'green' questions apart, as if the lexical similarity between them were a simple accident, perhaps it is more productive to try to bring them together. As we describe in the following chapters, when people in Macclesfield nowadays talk about litter, graffiti, speeding and bad parking, the kids hanging out in the underpass or in the park, shopping trolleys in the river and many other mundane annoyances, they are certainly giving voice to concerns that are perfectly familiar to environmental criminologists. They are interested in having these things cleaned up so they don't just get worse. They are also concerned about whether driving and parking can be made safer and better regulated, and with the ways housing developments impinge on existing places and spaces, both while they are being built and subsequently. But they also know that their observations of these things are the local instances of wider changes in the world. Such observations are not, as fear-of-crime theorists (including the present authors!) have often claimed, 'metaphors'. Rather, they stand in a metonymic chain of associations with the wider phenomena of which they are a part. This litter here, this excessive traffic, this pollution in the river, this egregious and unwelcome development on an environmentally sensitive site is replicated elsewhere and form part of the same systemic dynamics.

For these reasons, we want to pay particular attention in this book to people's responses to these challenges, whether those involve a sense of impotence and resignation, or are seen as a prompt to engagement and an effort to express care for the future and to repair damage in the present. Those living in Macclesfield experience these dilemmas in much the same way as their counterparts in other towns and cities around Britain, or Europe and perhaps beyond. The ways in which they navigate the claims and pressures of environmental change and degradation, the oscillations in public discourse between the sense of emergency and disabling fatalism, are pretty much the same invidious choices that the present conditions impose on us all.

Both environmental criminology and urban studies have long been concerned with the aesthetics and meanings of order and disorder, and the attendant vulnerabilities and injustices of the 'in place/out of place', in everyday urban life (Cresswell, 1996). But they tend to approach these issues in very different ways. In urban studies and allied disciplines urban green spaces, for example, have increasingly been described as safe and therapeutic landscapes (Marsh et al, 2021) – 'micro-geographies of healing and respite', as Doughty et al (2022: 661) put it. Such spaces have even been shown to reduce crime and disorder across a range of national contexts (Wo and Rogers, 2024). At the same time, urban green spaces are purported settings for the visible disorders of environmental criminology (Taylor et al, 2019: 13). Green spaces have become a site of ambiguity and contention, caught between claims of their therapeutic benefits and concerns over the reproduction of urban insecurities and inequalities (Pearsall, 2018; Kocisky, 2022).

In trying to both recover and perhaps extend the meaning of terms such as 'ecology', and its uses in the pasts and present of social studies of crime and social ordering, we want to explore a range of connections between people's articulations of their concerns about crime and harm here and now and their interests in the current condition and possible futures of the wider world. The modest troubles that people in Macclesfield bring forward for attention – the littered places, the inconsiderate driving and parking, the ecologically questionable planning applications – are instances in a minor key of much more menacing and far-reaching harms and risks. To interpret these connections better it seems helpful to encourage dialogue between criminologies centred in place (the focus of ecologies of crime and disorder down the decades) and the environmental and political concerns of the green criminological tradition.

TWO

New Ecologies of Security

Our study examined the ways in which people in one middle-sized town in north-west England mobilize ideas about harms or threats to their local environment (litter, mess, damage and so on) to think about questions of safety and security. While residents themselves often note that these concerns are of small scale, and even downplay them as trivial or parochial, they sometimes also lead to wider considerations about guardianship, care and responsibility for common environments. These observations prompt reflection on the uses of ideas like 'environment' and 'ecology' in criminological discussion, as just discussed in Chapter One. There are indications of a 'green' inflection to people's ways of voicing their concerns about local troubles in ways that make them distinct from our prior experience of talk about crime and anti-social behaviour in the same place, a quarter of a century before (Girling et al, 2000). When people in Macclesfield discuss and worry about local disorder nowadays, they see it less as a harbinger of property or violent crime and, rather more, as a sign of environmental degradation, and the erosion and neglect of shared amenities and public spaces. These are cues of disorders, and indications of harm, not necessarily associated with 'bad actors' and external intrusion, but with collective behaviours and, indeed, the actions of the very people experiencing them. They also implicate questions of state inattention and neglect in new ways. We turn to these matters in Chapter Three.

People generally continue, then, to be troubled by the traditional concerns of environmental criminologies in the form of low-level disorders and incivilities. But they also draw connections between some of these problems – traffic, litter, signs of drug use – and more fundamental threats to the environment and hence to our common security (Harrington and Shearing, 2017). For this reason, a new interpretive framework is required for understanding people's responses to problems that arise in their local environment; one that bridges the foci of environmental and green criminology, and connects concerns with everyday disorder to wider questions of how to orient oneself to security in the Anthropocene (Holley and Shearing, 2018), or what Thaler (2022: 13–27) calls 'a climate-changed world'. Such problems matter as much as they do nowadays because they are entangled with perplexing questions about who should act, and how each of us should act, in the face of threats that erode the boundary between the very local and the planetary and which partly de-centre crime within discussions of local disorders and harm.

This chapter aims to develop this framework – to sketch the outlines of what we might call a 'new ecology of security' or, to use a stricter but uglier formulation, to ecologize security. To do so is, we shall show, to acknowledge and work with the 'interplay between humans, landscapes and the non-human' (Harrington and Shearing, 2017: 59) in creating and sustaining conditions for ongoing co-existence, thereby extending the range of actors and actants under consideration when we examine what it means and takes to live collectively and comfortably with risk (Loader and Walker, 2007). It is to problematize the notion that being and feeling secure in everyday life is intimately tied up with freedom from concern and to propose that relations of care have today been rendered central to prospects for creating safer (and better) socio-natural landscapes. It also entails giving up on the idea that the environment is a silent backdrop to conflicts over human in/security and, instead, re-positioning 'nature and humanity not as separate but as entangled, plural entities embedded in intimate and constitutive relations with each other' (Harrington and Shearing, 2017: 65).

Placing green criminology

Green criminology seems like a good place to start in our search for this new ecology. Since its inception as an explicit project in the 1990s (Lynch, 1995; South, 1998; see also Goyes and South, 2017), this sub-field has developed into an active, innovative and diverse body of theory and research (Brisman and South, 2020). But it has forged theoretical orientations, objects of study, and political and policy outlooks that appear markedly different from those that have constituted environmental criminology, or what green criminologists sometimes demarcate as 'place-based criminology' (White, 2008). There have, to be sure, been occasional points of fruitful cross-over, such as Lynch and Boggess's (2015) analysis of the intersections between social disorganization theory and research on eco-cities. But with the exception of litter and fly-tipping (Groombridge, 2013; Johnson et al, 2020), the harms that preoccupy green criminology tend to remain outside of the standard agenda for tackling 'disorder' or 'anti-social behaviour'. Conversely, the 'disorders' that typically feature in environmental criminology tend not to be interpreted within a green criminological frame, nor acted upon as ecological problems.

Green criminology has been defined as the 'study of crime, harm and injustice related to the environment and to species other than humans' (South, 2014: 7; see also White, 2022). It is focused on the causes, consequences and prevalence of environmental crime and harm; legal and regulatory efforts to prevent and control such crimes and harms, and on how environmental threats are mediated and culturally represented (Ferrell, 2013; Brisman and South, 2014). As such, green criminology has made numerous significant contributions.

First, green scholars have drawn attention to – and developed our understanding of – a wide range of often serious, but still often overlooked, threats to the environment (White, 2014). These have included, *inter alia*, deforestation, toxic waste, wildlife destruction, and the cognate harms of fossil fuel and other extractive industries; threats to the liveability of urban environments from fire, flooding, or water and air pollution (Walters, 2014; Brisman et al, 2018); animal abuse, including the legal and illegal trade in non-human animals, and questions of food production and safety; as well as a range of crimes and harms stemming directly or indirectly from global warming and climate breakdown – as catalogued in Rob White's (2020) *Climate Change Criminology*.

Second, green criminology has been at the forefront of contemporary efforts to ask hard questions about the relation between 'crimes' and 'harms'. The analysis of 'social harm' – or zemiology (Canning and Tombs, 2021) – has emerged both within and in opposition to criminology, its point of departure from criminology's focus on state-sanctioned behaviour being that the some of the most systemically harmful acts in capitalist societies are not proscribed by criminal law (Pemberton, 2016). Environmental damage has become a key site for making and developing this point. Much green criminological research does focus on acts that are criminalized, its point often being to expose and document the neglect and under-enforcement of what are often forms of serious offending. But green criminology has also demonstrated time and again that 'many conventional, and legal, forms of human production and interaction do far worse things to the natural environment than those activities deemed illegal' (White, 2008: 18). Many of these crimes and harms flow from a powerful state/corporate/colonial nexus and are under-enforced or not criminalized precisely because criminal sanction and enforcement would threaten entrenched political and economic interests (Lynch and Long, 2022). Sometimes this has prompted arguments for criminalization, as in debates about the crime of ecocide (for example, White, 2018). But green criminologists have also shown how environmental harm is bound-up with taken-for-granted – even prized – aspects of modern forms of life, especially in the Global North, and have focused on what Agnew (2020) calls 'ordinary harms': acts 'performed widely and regularly by individuals as part of their routine activities [which] are generally viewed as acceptable, even desirable and collectively have a substantial impact on environmental problems' (2020: 52). We turn to one of these acts – mass motoring – in Chapter Five.

Green criminology has since its inception understood itself as a form of critical criminology. It has, as such, been pursued with an alertness to crimes of the powerful, and to questions of power, injustice and inequality. It has in this respect, thirdly, brought to criminological discussion a philosophical attitude that has problematized the distinction between 'society' and 'nature' and – to varying degrees – de-centred the human, and it has coupled this with a concern to extend what we consider to be at stake in contests over justice. Differences of perspective can be found here between the competing claims of anthropocentrism, bio-centrism and eco-centrism and the contending visions of environmental, ecological and species justice (White, 2014; Brisman and South, 2018). The key takeaway – or at least the one we wish to develop in this book – is an expansion of our field of vision beyond human-to-human harm and disorder to also encompass harms inflicted by humans on non-humans and by non-human entities on human beings (Holley and Shearing, 2018: 11).

Given these contributions, green criminological perspectives offer vital resources in our effort to understand the meanings and effects of everyday disorder in a climate-changed world. But there are also grounds for thinking that green criminology remains – on its own – insufficient for this task, at least as we have encountered it in this study. We need to reconfigure its orientations in a couple of ways.

Green criminology has long been attuned to questions of scale, and at least partially to matters of place. Environmental harms are typically transnational in character and have no respect for national borders built by humans. Think of polluted air or toxic rivers. It is also commonly recognized that while environmental crimes tend to be global in kind, their impacts are always also local *somewhere*, albeit in sharply differentiated ways. Green criminology is focally concerned with threats to, and the protection of, the habitats that sustain human and non-human life (Goyes et al, 2021). Recent developments in Southern green criminology have also brought to the fore the importance of indigenous ways of seeing and feeling 'place' and attendant understandings of human entanglement with the land (Goyes, 2019; see also Bawaka Country et al, 2016; Yarwood et al, 2025).

But, as mentioned, green criminology has its intellectual and political roots in radical criminology, not in the socio-spatial analysis of crime and victimization, nor in urban studies more broadly. As a result, questions of how environmental crimes and harms are connected to the changing character of urban spaces or mediated by people's situatedness in – and sense of attachment to, or disconnection from – their street, neighbourhood, town or city have not always been focal ones for it, at least in terms of its larger theoretical contours.

Green criminology's relation to place has, in these senses, tended to be 'thin' rather than 'thick'. In practice, when researchers undertake empirical work on environmental harms, they necessarily start to explore their impacts in particular places. A collection of case studies on *Green Crime in Mexico* (Arroyo-Quiroz and Wyatt, 2018), or a study of the local politics of an industrial plant closure in North-East England (Davies, 2014) are excellent examples of work of this kind, as is Lundberg's (2022) account of environmental harm at the 'edges of the city'. What is more doubtful is the degree to which such studies have yet been understood as potential meeting points for green and urban criminology, at least within the criminological tradition of 'ecological' approaches. This is how it comes about that green criminologists seldom believe they have anything distinctive to contribute to debates on 'the problem of street crime and its control' (Lynch and Boggess, 2015: 322).

Our hope is that a more convergent discussion of 'ecologies' offers an invitation to green criminology to attend more closely to everyday socio-natural 'harmscapes' (Berg and Shearing, 2018) beset by a confluence of climate and other disorders, albeit in ways differentially embedded in the experience of particular places. But a complementary extension of socio-spatial criminology is also called for here. It can fruitfully meet green criminology halfway, as it were, by giving up on the idea of urban environments as the mere backdrop to human-centred crime and disorder and, instead, see the city as a tangle of machines and infrastructures, humans and non-humans, matter and nature (Amin and Thrift, 2016), while approaching urban environments as sites of ecological worth (Gandy, 2021). That is the altered lens through which we propose to think about everyday disorder.

As a field of study, green criminology has tended to operate with – even be constituted by – a sharply delineated focus on 'environmental' crimes and harms, which its practitioners treat as a distinct and neglected sub-set of larger legal and social categories. It has, as mentioned, directed much analytic effort towards problematizing the distinction between 'crimes' against the environment and environmental 'harm' – to beneficial effect. But along the way, less attention has been devoted to specifying the properties that make a crime or harm 'environmental' and what demarcates these acts from other offences and troubles. The result is an inventory of crimes/harms that green criminologists study: pollution, wildlife destruction, deforestation and so on; and a large number of crimes/harms that are left to the rest of the field: robbery, burglary, sexual violence, criminal damage, fraud, bike theft, and so on. But are these demarcations really so sharp? Suppose that the fear or experience of robbery, or of having one's bike stolen, leads people to get around by car – rather than by walking, cycling or taking public transit – with all the environmental harm that this entails. Does that make robbery or bike theft – however indirectly – a crime against the environment? Some green criminologists have certainly been very alert to this issue. Lynch and Boggess's (2015) attempt to 'green' social disorganization theory is a case in point. In a cognate vein, Brisman (2004) has analysed the damaging environmental effects of restrictive re-entry policies for ex-offenders in the United States, detailing how they undermine sustainable communities; in a related paper he outlines why we should consider criminal disenfranchisement an environmental issue (Brisman, 2007).

But if we are to deploy the insights of green criminology to ecologize questions of security we need to press this point further. This means softening the distinction between crimes/harms against the environment and other offences, and perhaps even letting go of 'green crime' as a discrete category. Instead, we might preferably treat 'environmental' as a way of seeing rather than the thing that we see – a lens on a wide canvas of crimes, harms and disorders that helps sharpen our sense of the damage caused, why these harms matter, and how one might best respond to them. We need, in other words, to treat the environment 'less' as 'an object than a point of view' (Charbonnier, 2021: 16).

We might also extend this insight to green criminology itself (though given we have to date worked as sympathetic outsiders to the green criminological tradition, we make the point respectfully). Over the last three decades, green criminology has become a thriving and diverse field of work – replete with its own internal differentiations and labels (see Sollund, 2015: 4; White, 2022: 20–2) – while arguably exercising 'marginal influence' over the remainder of the criminological field, whose practitioners have pressed on largely unchanged (Harrington and Shearing, 2017: 146; Lynch and Long, 2022), believing that not only environmental crimes/harms, but the entire ecological question, has been 'outsourced' to 'the environmentalists' (Klein, 2014: 7). Against that backdrop, perhaps what a new ecology of security needs is not so much a distinct green criminology as the recovery of that field's pioneering promise: one that treats 'thinking green as a sensitizing *perspective*' (South, 1998: 212; emphasis in original) and carries an ambition *to green* how we generally approach questions of disorder and security.

Security in a minor key

In *Why Things Matter to People*, Andrew Sayer argues that our relationship to the world is not just one of perception or stimulus, but of concern. We care about the world we inhabit, act in it, evaluate it, and are vulnerable to it. We can flourish or suffer in it (Sayer, 2011: 3; see also Rosa, 2019: 126). Thinking carefully about these aspects of people's relations to the settings of their daily lives, Sayer argues, affects the kind of social science we practice, because it draws our attention to the duality of capability and vulnerability that characterizes our environment of action:

> How people can best live together is not merely a matter of coordination of the actions of different individuals by means of conventions, like deciding which side of the road to drive on, but a matter of considering people's capacities for flourishing and susceptibilities to harm and suffering. When we think about how to act, we do so with some awareness of the implications for well-being, both ours and that of others. (Sayer, 2011: 8)

Sayer's remarks are important for the revised ecology of security we wish to develop, especially when we think about people's ways of inhabiting a specific socio-natural landscape. We might begin from a premise that people feel quite strong concerns over how their place is faring in the world, whether they feel it to be rising or declining, or just about holding its own. They might feel that it is a relatively safe and protective space, though the world beyond is more dangerous, or indeed vice versa. They might be concerned not just about whether they, or their partners or children, are safe while out on the street during the hours of darkness (the classic fear of crime prompt question), but whether they are in some other way at risk, from what they are receiving – or sharing – through their computer screens, to the looming dread of local flooding or wider environmental catastrophe. These are matters of concern, and the ways in which they are so are worth exploring, without presuming very much about the objects of people's concerns, nor how intensely they feel about them.

By contrast, one of the attributes of people's conduct that most exercises some of our participants in this work is what they interpret as evidence of a *lack* of concern. That may be a feature of individuals' actions when they drive or park inconsiderately, or when they drop litter in the park. Yet it also applies to the acts and omissions of corporate actors and people in positions of authority when they make planning decisions or fail to provide enough resources and attention to maintain valued local amenities and spaces. Such omissions may undermine people's sense of security in the continuity and dependability of their lifeworld. They may also indicate a failure to show concern for the future in ways that inherently invoke questions about environments under threat of deterioration.

There are many registers of concern that may be brought to mind when we discuss questions of safety and security, some of which more forcefully claim our attention than others. In *Lively Cities*, Maan Burua speaks of urban theory 'in a major key', by which he means: 'those modes of thought and practice that define what the primary matters of concern are, and that wield influence in dictating what relations and events are worthy of conceptual intervention and inquiry' (Burua, 2023: 2). 'Security' is also framed, promised, pursued, discussed, and critiqued 'in a major key'. Very often, these 'major' framings involve questions of national security, or security against threats such as terrorism. There are, nowadays, climate change-informed variants of this major framing of security (see, for example, Busby, 2022).[1] These can spill over into mundane social life insofar as measures taken against them – in the form of enhanced police powers, or the general adoption of methods of surveillance or border control – condition our relations with state institutions, and general sense of the dangerousness of the wider world. Beyond that, security has commonly come to reference criminal threats to people's person and property, authored by dangerous others who intend deliberate harm. This framing pits a 'law-abiding majority' against a 'criminal minority' from whom the former needs protecting – a minority that it typically 'othered' along lines of ethnicity, age, gender, class, nationality and religion. What follows is a familiar repertoire of institutional response, typically focussed on police (more officers, with greater powers) and resolute punishment (which conventionally means prison). This is the security imaginary that tends to dominate public discourse, as well as supplying the staple of criminological discussion.

Many critical analyses of security take issue with aspects of this 'major' framing. Scholars of securitization, for example, are concerned with the often-insidious nature of security claims and their effects upon contemporary society and culture (for example, Schuilenberg, 2015). Security is conceptualized here as a set of political practices; governmental speech acts mobilized to lift 'security' from the constraints of normal democratic politics and justify decisive, exceptional measures in the face of what is presented as an existential threat (Balzacq et al, 2016). A radical variant of the same suspicion pervades critical security studies (Neocleous, 2008), as well as an emergent anthropology of security (Goldstein, 2010). Security, on this view, is a set of powers and techniques that are marshalled in the name of the state against minorities. Security politics has a predilection for urgent and speedy action that is inimical to due process and democratic deliberation. It works with a conception of political community founded on a sharp friend/enemy distinction. Given this, the key analytic task is unmasking and critique. The attendant political ambition is to transcend security as a frame for thinking about safe and viable political communities. Security cannot be redeemed or reframed in a 'minor key': it has to be replaced.

Some other critical analyses of security have taken a different path, that of recovery and re-interpretation. One such account takes issue with the reductive association between security and physical threats to persons and property. The resultant task is to expand the meaning of security to include adequate satisfaction of a wider range of basic needs, such as food, shelter, and housing (Paris, 2001; Starke et al, 2025). As the United Nations Development Programme report that initiated the idea of 'human security' put it back in 1994: 'A feeling of insecurity arises more from worries about daily life than dread of a cataclysmic event. Job security, health security, environmental security, security from crime – these are the emerging concerns of human security all over the world' (1994: 3). On this view, security is not an idea to be critiqued and superseded but expanded to encompass a wider ambit of social harms. Security becomes the lens through which to approach and take appropriately urgent action with respect to a whole raft of public problems.

In *Civilizing Security*, Loader and Walker (2007) offer an alternative attempt to reconfigure the theory and practice of security. They start from the premise that individual security has both objective and subjective dimensions and refers to material and psychological states of (well) being, and thus to the questions of flourishing and suffering that Sayer raises. Loader and Walker seek to identify the properties and virtue of security understood as a thick public good. Security has to do with creating institutional arrangements that foster and sustain people's sense of effortless belonging and enable all members of a political community to live collectively and comfortably with risk. On this view, being and feeling secure is, in an important respect, to be free from routinely having to attend to one's safety, and in this sense to be removed from a condition of care (see also Hamilton, 2013). But is this version of security as freedom from care fit for a time of socio-ecological crisis? Is being care-free precisely *not* the kind of ethos that is called for and upon if we are to foster anything approaching security in a climate-changed world?

These efforts at humanizing and civilizing security are attempts to re-cast the matters at stake using 'minor theory', to borrow Cindy Katz's (1996) term. They offer 'different treatments of the same language' (Burua, 2023: 32) in order to supply resources for thinking and acting otherwise about the meaning and practice of security. A new ecology of security needs, however, to go further in this regard. Burua makes a case for practising 'urban theory in a minor key' by conceiving the urban as an ecological formation – living fabrics of association between human and more-than-human forces. In so doing, he writes: 'The minor exists only in relation to the major; the two cannot be decoupled. However, the minor can reveal an entirely different story stirring within any field. … The minor breaks from conventional archetypes and established associations of thought practice' (Burua, 2023: 15).

To ecologize security is to re-play it in a minor key in precisely this sense: the aim is to disclose and develop an 'entirely different story' lurking in dominant framings and critiques of security. That story has to connect with the enduring austerity that now marks British economic and political life, and to climate breakdown, in ways that enable alternative thinking about the referents, threats, agents, means and ethos implicated in the question of what it might mean to be and feel secure (McDonald, 2021: 2). It is a framing of the security question that revisits the 'old' threats that dominate security in a major key and asks afresh about their import and meaning, while making room for disorderly and harmful objects that major theory expels from consideration as security concerns. It acknowledges the malign threats to safety that continue to emanate from 'the outside', while being open to the ways in which 'we' are today implicated as contributing authors to the security concerns that so trouble us. It recognizes, while seeking to reconsider and revise, the role and purpose of standard security institutions (notably, the police), while asking new questions about who is responsible for safety and security today and what it means to discharge that responsibility in a climate-changed world. It remains alive to the political re-alignments and possibilities that emerge from these 'minor' modes of interpretation and diagnosis.

In broad terms, the revised ecology of security we propose makes two conceptual moves. Borrowing terminology from Sarah Whatmore (2002), we might call these 'embodying' security in the everyday and 'enlarging' the set of feelings and affordances required for us to inhabit safer socio-natural environments.

Fassin (2013: 621) argues that 'attention to the ordinary' provides 'a way to transform our perspective on the world'. A new ecology participates in a shift towards thinking about security concerns as being situated in the everyday life of the very different places in which they arise. This interest is a point of continuity with our earlier study of people's fears and feelings towards crime in Macclesfield (Girling et al, 2000). Yet the conditions of 'the ordinary' may themselves undergo far-reaching change illuminating faultlines in the taken for granted uneventfulness of everyday life. This may be because the socio-political-ecological contexts that condition prospects for secure life in any given place have undergone – perhaps dramatic – alteration. It may be because the infrastructures of everyday security can change quite rapidly. Who provides what forms of security, against which kinds of threat, using what kinds of technology? It is vital that we seek to understand these matters in their 'molecular' aspect (Schuilenberg, 2015: xx) though close, contextual observation and enquiry.

Considerations of people's 'mundane, material embeddedness in the world' (Felski, 2000: 79) provide a way in to thinking about questions of safety, danger and security in a minor key.

They remind us why people's experience of place and locality remain of utmost significance, and why the sense of living under safer or more dangerous conditions there is of central importance to them. Emergent literatures in a number of fields increasingly investigate themes of 'everyday security', including the ways in which differently situated people, including those in extremely difficult and dangerous situations (Ajil et al, 2020), try to construct and sustain passably secure daily lives (Jarvis, 2019; Nyman, 2021).

Crawford and Hutchinson argue that in order to capture adequately the 'quotidian aspects' of these matters, there is a need to focus upon 'security experiences'. In the first place, these concern: 'the lived realities of practical security measures, including the diverse ways in which programmes, strategies and techniques for governing security are experienced, taken up, resisted, and even augmented by different individuals and groups within society' (Crawford and Hutchinson, 2016: 1185). But the analysis and critique of security, Crawford and Hutchinson argue, has fallen overwhelmingly on the side of programmes or 'logics' of government, rather than that of ordinary experience: it has been locked – in our terms – in a major key. One result of this preference has been a limited grasp of what it means to live under particular regimes or dispositions of 'security'. For this reason, Crawford and Hutchinson suggest:

> Asking questions therefore about the everyday lived experiences of security – and not just how projects are 'implemented' on the ground, but how they are felt, lived through, sensed and borne by individuals and groups – is vital for understanding security governance writ large. Yet such questions have been largely absent from contemporary security research. (Crawford and Hutchinson, 2016: 1186)

This refocusing on the 'messy everyday world' (Crawford and Hutchinson, 2016) of security proposes connections to the kinds of topics that Felski (2000) indicates are integral to the social analysis of everyday life in general – our routine uses of time, the centrality of home, the meanings of dwelling in specific locales, and the affective/emotional textures of those practices and experiences. Such an approach 'asks us to consider the more mundane, ordinary routines and day-to-day discussions and practices that people engage in to help manage their own safety'. 'It is in part through such everyday security processes', Crawford and Hutchinson (2016: 1188) suggest, 'that ordinary people foster security for themselves and for others while striving to live with *in*security'.

This orientation to 'the ordinary' is a key ingredient of the analysis of harm and disorder we present here. But we need, further, to ecologize security by enlarging our sense of what is encompassed and at stake in seeking to create and sustain safe operating space for human and more-than-human life. The now familiar concept of the Anthropocene (coined by scientists Paul Crutzen and Eugene Stoermer in 2000) expresses the idea that humans have become geological actors, fundamentally altering the planet at an increasing and alarming rate (Steffen et al, 2015), giving rise to what Latour (2018) calls 'a new climatic regime' (Harrington and Shearing, 2017: 26–31; Matthews, 2023: ch 1). In *Security in the Anthropocene*, Harrington and Shearing (2017) intimate a number of ways in which this mutation impacts how we think about and act upon questions of security.

It requires, first, that we abandon the idea that humans 'live *on* rather than *in* or as *part of* the earth' (Harrington and Shearing, 2017: 38; emphasis in original). This entails that we stop treating the environment as the silent, inert backdrop against which the local and global drama of human in/security unfolds and reckon, instead, with the 'intrusion of Gaia' into 'the heart of our lives' (Stengers, 2015: 47; see also Latour, 2017: 75–110). This necessitates recognition of human embeddedness in ecological processes of which we are each an integral part and on which we all depend. We must,

Harrington and Shearing (2017: 52) argue, 'contemplate a security ecology incorporating multiple, entangled actants'. If we view 'humans as intrinsically linked (though not bound) to other modes of being, including organisms, animals, machines, feedback systems etc.', they continue, '[we] can open up new understandings of power, sovereignty, responsibility and harm as well as new, immanent possibilities for action' (Harrington and Shearing, 2017: 33).

We need, secondly, to revise our thinking about the referents of security, and how we understand and investigate relations of threat and vulnerability. This means, as mentioned, extending analysis of crimes, harms and disorder from human-to-human encounters to those involving humans-to-things and things-to-humans. But it also means unsettling distinctions and mental schemas 'that are a product of an age we are careening out of and cannot be so easily translated to the new epoch' (Harrington and Shearing, 2017: 87). To be sure, people and places will continue to be troubled by 'external' criminal threats, and demand modes of protective defence from them. But a new ecology of security needs to problematize the us/them, friend/enemy, inside/outside, norm/exception couplets that characterize security in a major key, and attend as well, or maybe even instead, to what Harrington and Shearing (2017: 122) call the 'relentless insecurities of the unexceptional'. This is likely to entail developing accounts of distributed agency and responsibility that recognize the many ways in which 'we' have become implicated in the everyday security problems that so concern us. It makes little sense to wage 'war' in the name of security if that means a fight to defeat ourselves.

It follows, thirdly, that we need to reconfigure the relation of security to care. There is an influential strand of theorizing that – drawing on its philological and historical roots – portrays security as a state of being and feeling free from care (Hamilton, 2013). But under conditions of durable austerity and profound ecological mutation, we might instead suggest that sustained carelessness has become an agent of harm and disorder. To be carefree denotes a condition of indifference towards, and separation from, others, human and non-human. Drawing on a body of work on care ethics (for example, Tronto, 1993), Harrington and Shearing sketch an alternate ethos of security grounded *on* notions of care. Security, on this view, flows out of particular and concrete relationships; its rests on a situated, cautious and worldly 'awareness of entanglement and relationality' (Harrington and Shearing, 2017: 131); it emphasizes local solutions to specific problems while remaining attuned to relations with multiple life-worlds: to distant places, and more-than-human beings and forces (Harrington and Shearing, 2017: 117). It is a mode of ecological security characterized by McDonald (2021: 95) as 'a commitment to the resilience of ecosystems themselves and, with it, the rights and needs of vulnerable being across time (future generations), space (impoverished populations), and species (other living beings)'.

Attending to everyday socio-natural harmscapes

Towards the end of *Security in the Anthropocene*, Harrington and Shearing (2017: 147) worry that 'readers may be frustrated by the lack of "real world" empirical examples'. We were not, it should be said, among such readers. But we nonetheless hope and intend that what follows in the rest of this book is a grounded analysis of everyday disorders that is broadly in keeping with, and seeks to extend empirically, the approach to security that Harrigton and Shearing outline.

In this respect, the minor conception of security we have sketched is not simply a theoretical exercise. It has been developed with a view to nourishing the connections 'between analytical adventure and everyday apprehension' (Whatmore, 2002: 176) of concerns about the world and its condition; as a means of seeing and grasping what is going on in 'the most mundane of worldly transactions' (Whatmore, 2002: 177). In a cognate vein, Les Back (2015: 832) argues that focusing on the everyday means attending closely 'to what is easily discarded as unimportant'. By so doing, Back suggests, it becomes possible to 'identify public issues in the mundane aspects of everyday life' and 'link the smallest story to the largest social transformation' (Back, 2015: 834).

Doing so demands two things of us, and perhaps of our readers. First, it requires that we conceptualize the town where we conducted our research not simply as a place, still less as a community (in the tradition of community studies – cf Crow, 2018), but as an everyday socio-natural 'harmscape' (Berg and Shearing, 2018). This means approaching it as an entity composed of human and non-human life, objects and infrastructures, built and physical landscapes, and treating the harms and disorders found in its homes, streets and neighbourhoods as having something to do with relations between these entities. The precise nature of that 'something' is a matter for empirical exploration.

Second, it requires cultivating and practising what Anna Tsing (2015: ch 1) calls 'the arts of noticing' (see also Stengers [2015: 62] on 'paying attention'). This means observing closely, and listening attentively, to what people tell us about the kinds of troubles that pervade their ordinary lives, without preconceived criminological ideas about what these troubles are supposed to be and how we should rank their seriousness. And it means bringing appropriate analytical tools and ways of thinking to bear on the connections that people make – and those that we might seek to formulate – between the local troubles our participants describe and the present condition and anticipated futures of the wider world.

THREE

Socio-Natural Harmscapes

People who live in Macclesfield generally feel positive about it. Many are attracted to the town, and/or actively choose to stay and make their lives there, by both its attributes – a 'good social mix', 'friendliness' and 'strong sense of community', and good schools – and by the relative absence of the social problems that attend life in large cities. Macclesfield is not very ethnically diverse. According to the 2021 Census, 94.1 per cent of the town are white, 2.4 per cent Asian, 0.6 per cent black, and 1.8 per cent are mixed or from other ethnic groups. It may be that for some people in the town this homogeneity contributes to their sense of comfort with the place. But some of our participants also coded the lack of diversity as a sign of Macclesfield's 'boringness', as if it has fallen out of step with progress in the wider world. As one female resident put it: 'I don't think there's any big stressors for me here. There's nothing that really aggravates me. Okay, it's not the most exciting place in the world, but equally, it's not overly stressful either.' Another female resident made a similar observation in more emphatic terms:

> It's the friendliness, the people. It's small enough that people know each other. They know me, they know my children, so I always feel that they are quite safe because, it's walking distance so I can let them go out now they're teenagers. They can go about town, and I let them walk home and things because it is mostly walking distance. So, we let them do that. For me, I commute

> into Manchester for work so that's a good link, I like that. And the hills, nice, next to nature. It's nice you can get out and feel you're in the countryside very quickly. So yes, all of those things.

Prominent among the town's felt benefits were its accessibility, its connections with other big urban centres such as Manchester (and its airport), London and everywhere else. These links have enabled Macclesfield to attract and retain the pharmaceutical industry that helped this former 'silk town' to avoid the worst ravages of decline that afflicted other northern towns in the late 20th century (Niven, 2023). They allowed it to modernize, restructure, survive and modestly prosper. At the same time, many of those drawn to move or return here strongly desired the town to retain aspects of its 'northern-ness', its industrial and craft heritage, its homeliness, and particularly its rural surroundings.

The accessibility of 'wilderness' and 'nature' were prominent in accounts of Macclesfield's attractiveness as a place – feelings that were brought into heightened focus during and in the immediate aftermath of the COVID-19 pandemic. Residents often spoke of seeking and finding a lifestyle 'close to nature' in a place 'where very suddenly, you're in this really nice kind of wilderness' (male resident). For many, the town is perceived and enjoyed as what Angelo (2019: 648) calls a 'socio-natural' environment – a place where 'nature' doesn't simply exist 'out there' but is experienced 'in here' as an integral feature of the built landscape. Views beyond the town are encountered while shopping in the town centre or turning down many residential streets. As one female participant in a town centre focus group remarked: 'that's one of the things I like about Macclesfield, that it feels like you're sitting in hills'. This perception is clearly part of what animates the various forms of environmental activism one encounters in the town. It may also serve as a pointer to how the (local) environment, and its troubles, are more generally apprehended and acted upon.

The forms and meanings of chronic harm

How does worry about crime and disorder fit into this sense of place? People living in Macclesfield tend to think of it as a relatively safe place. Yet there were significant pockets of concern, and some indications of growing uncertainty and unease. In our 2021 community survey, 53 per cent thought Macclesfield was safer than elsewhere in the UK, while 39 per cent thought it about the same. In 2022, 30 per cent thought it safer than elsewhere in the UK, with 50 per cent saying about average. Levels of victimization reported in the survey were correspondingly low, 4 per cent 'since the pandemic began' in the summer of 2021, 7 per cent 'in the last year' in 2022. In the earlier year, however, 29 per cent said they would feel 'fairly' or 'very unsafe' walking in their area after dark; this grew to 42 per cent in 2022. Similarly, while in 2021, 80 per cent felt their life 'these days' was very or fairly secure, only 56 per cent said the same in 2022 (and the proportion responding 'it's difficult to say' grew from 5 per cent to 31 per cent over the same period). The crime issues that most concerned residents in 2022 were burglary (32 per cent fairly or very worried, compared with 28 per cent in 2021) and online fraud (also 32 per cent in 2022; but 44 per cent in 2021). Some 21 per cent were worried about being a victim of violent crime in 2022 (12 per cent in 2021), while 20 per cent were worried about being mugged (12 per cent).

Accounts of Macclesfield as a (relatively) safe place were also prevalent in exchanges with residents:

> Empty streets in cities have made me … a bit nervous, at night. … Or not, maybe not nervous, isn't the right word, sort of cautious, and I would, if I was walking home, when there was nobody on the street. … I remember being on a 'Reclaim the night' march in Liverpool in … um … early 80s, I guess? But I have never really felt nervous in Macclesfield. Even though there are deserted

> street[s]. It almost makes it feels safer. (Audio diary, female resident, 11 November 2020)

When asking residents about what troubled them about the places around their homes and neighbourhoods, crime-related experiences and insecurities were rarely foregrounded. The overall sense of our findings is that crime has become an *episodic* feature of most people's sense of place – a relatively infrequent occurrence that may happen *here* because it sometimes happens *everywhere*. Even those female respondents who built attentiveness to crime risk into their daily routines (being alert to their surroundings, avoiding places like underpasses and canal paths, especially after dark) typically remarked that such risks are present here because women encounter them everywhere (Miles et al, 2026). This compares strikingly with our earlier study in the mid-1990s, when worries about crime – especially burglary and car crime – were more pervasive, and central to people's sense of Macclesfield as a town that was targeted by 'travelling criminals' and in need of protection from incursion (Girling et al, 2000: ch 3). Furthermore, some offences that people do report being worried about, such as online fraud, do not feature prominently in people's perceptions of place – they are non-discursive crimes, forms of victimization seemingly disconnected from place and difficult to frame as a public issue. Local place-talk in Macclesfield today is not first and foremost talk about crime.

The same cannot so clearly be said of residents' concerns about everyday safety more broadly defined, and about low-level 'disorder', 'anti-social behaviour' and related issues. Our findings suggest that harms to the local environment (whether criminalized or not) loom large in people's concerns about the town. If feelings of security mean being able to inhabit an environment in which one's life projects can proceed with some degree of certainty and predictability (Giddens, 1991; see further Loader et al, 2023), people may be particularly

Table 3.1: Neighbourhood problems: how much of a problem in your area is ...?

Percentage saying a 'fairly' or 'very' big problem	2021	2022
Drug dealing*	–	50
Badly parked cars	32	43
Speeding cars*	–	43
Rubbish or litter lying around	20	25
Homeless people living on the streets	13	23
People being drunk or rowdy in public places	15	16
Teenagers hanging around on the streets	13	16
Vandalism, graffiti and other deliberate damage to property	11	12
Noisy neighbours or loud parties	8	8

Note: *Question added in 2022 in response to themes emerging from qualitative fieldwork.

attuned to troubles that impinge upon a place in persistent and incremental ways rather than those that occur now and again. We term troubles of this kind *chronic harm*.

Table 3.1 shows the results from the two waves of the community survey. We asked whether a range of issues were considered by people to be problems in their neighbourhood. In 2022 fully 50 per cent felt drug dealing was a problem in their area (it was the first or second 'biggest' problem across all areas of Macclesfield), with 43 per cent saying the same about badly parked cars and speeding cars, and over 20 per cent indicating rubbish and litter and homeless people living on the streets. Concern about these harms seems to have risen between 2021 and 2022, and indeed across almost all the measures with data for both years.

Table 3.2 shows results relating to feelings of (un)safety. We asked, 'When you are using public spaces in Macclesfield,

Table 3.2: Neighbourhood safety: Do any of the following things ever make you feel unsafe?

Percentage responding 'quite' or 'very often'	2021	2022
Drug users/visible signs of drug use	17	31
Speeding cars*	–	28
The way some people drive or park their cars	20	28
The risk of being infected with COVID-19	23	20
People drinking alcohol in the streets	8	20
Young people hanging around	8	18
A lack of police presence	9	17

Note: *Question added in 2022 in response to themes emerging from qualitative fieldwork.

do any of the following things ever make you feel unsafe?' Concern again grew over the two survey waves across almost all the items shown, with signs of drug use, speeding cars, and the way people drive or park their cars eliciting the greatest sense of danger.

Some important shifts have taken place in public concerns about disorder since our study in the 1990s. Young people 'hanging around' in public places was a prominent (adult) anxiety in the town in the 1990s. So, too, was alcohol-related disorder in public space. These appear to rank much lower among the things that trouble people in the early 2020s. In the mid-1990s, cars featured in our study in two respects: people worried about their cars being damaged or stolen and complained, often indignantly, about the police targeting motorists rather than catching 'real' criminals (Girling et al, 2000: ch 6). One striking finding of our research today, by contrast, is that concern about cars features prominently on the list of public concerns: problems associated with car-centric mobility appear now to have a consequential impact on the felt safety and liveability of the town (Loader et al, 2025; see Chapter Five).

One of the persistent findings involved worries about drugs and the signs of visible drug use, though as we shall see, the meanings attached to this problem have changed over time in key respects. Today, these troubles loom large partly because of what they signify about the quality, liveability and sustainability of the everyday environment and the felt pervasiveness of such encounters. Next, we explore two such chronic harms in more detail – those that concern various forms of 'litter' (broadly understood) and those attached to the detritus of urban development. In each case, we reflect on the sensory intrusion of unwanted and troubling people, objects or behaviour upon local socio-natural harmscapes.

Littered harmscapes

In our surveys and interviews, and on social media, residents worried about inhabiting an environment that is today in various ways becoming frayed, degraded and spoiled. Taken one by one the problems here appear small, even trivial – potholes, neglected buildings, litter, graffiti, and so on. But they are not understood as small in their cumulative effects. This 'spoiling' or 'fraying' of the local environment is akin to Mary Douglas's famous description of dirt as an encompassing category of events or situations that 'blur, smudge, contradict, or otherwise confuse accepted classifications' (Douglas, 1975: 51). These pervasive effects on the felt safety and quality of the lived environment are laced with social meaning. Unlike online harms, the fraying of the material environment was a prominent and persistent means of talking about *this* particular place. But what is considered to be waste or trash, or litter, and what do its various forms signify?

Discarded paraphernalia of drug-taking and drinking appeared often in our participants' accounts of littering and litter-picking, coded as unsightly and unwanted reminders of troubling presences in everyday life (cf Murphy, 2012). The detritus of drug-taking was also typically read as a form of environmental pollution: 'You can smell marijuana a lot in the streets these days, all the time now', one male resident in his thirties observed. 'On the way here, I could smell it over there. So, people are smoking more flagrantly now, I suppose, aren't they?' Others registered it as a blight on the quality of daily life, as in this report to Fix-my-Street: 'There is smashed glass, needles, dog/human poo, graffiti. This alleyway is used daily by over 100 people. It's in an awful state.'

In 2020, a number of social media posts by Macclesfield Police drew public attention to small gas canisters as indicative of a drug problem and as a threat to young people in the town. 'Unfortunately', one officer reported, 'I have received several emails regarding nitrous oxide canisters (laughing gas) scattered around parks and streets' (Macclesfield South Police Facebook, 17 June 2020). In comments below these posts, residents reported seeing canisters littering their own areas outside local food stores, their children's schools and parks. When the posts were shared in neighbourhood Facebook groups, detailed comments supplied names of roads on which canisters had been seen. Members of the town's many litter-picking groups also reported seeing drugs detritus during their litter-picks: 'But as I say, you see the paraphernalia, you see the little pill bags, the popper bags, or whatever they're called. We've had a relatively large number of syringes behind the pavilion, but not so many recently' (male resident). Across our discussions, in fact, residents referred to seeing – or in some cases picking up – what they took to be 'evidence of drug stuff': 'needles', 'little foil wraps', 'little pill bags', 'popper bags', 'syringes behind the pavilion', 'empty gas canisters', 'broken glass bottles'.

These materials not only littered the local environment – making them in some circumstances indistinguishable from other debris of mundane consumption. They were also an uneasy reminder of hidden harm in the town's ostensibly safe environment, a vicarious encounter with another, more precarious Macclesfield. 'Unsavoury people can lurk in shrubs, can't they?', one litter-picker observed – inadvertently acknowledging that greenery can also hide illicit activity. 'I do quite a lot of work with a litter-picking group. And we probably collectively know where every drug den in Macclesfield is. Because quite a lot of drug users crawl into a bit of thick vegetation' (walking interview, male resident). Drugs detritus is not, then, simply a form of environmental degradation. It is also a disquieting symbol of criminality, vulnerability and exploitation, a surface indicator of 'of what's happening at the corners and the edges' of what it generally experienced as 'quite a safe' place (female resident). As one male resident reflected:

> You look behind the curtain of the underside of towns where you live, and in that sense, Macclesfield is really no different than anywhere else. It's not exempt from the social and economic problems that society faces, and I just happened to get a glimpse of it when normally I'm very lucky, I don't.

The natural beauty of the countryside that surrounds Macclesfield is typically the first thing that people mention when asked to describe the town. But the town itself was often characterized in these same conversations as unkempt and deteriorating – spoilt by the careless discards of other people's consumption that inscribes unwanted values on the place (Ablitt and James Smith, 2019). This too was a feature of our discussions, and a staple on social media. Participants often described the town as 'scruffy', a place where one sees 'lots of rubbish in the street'. People bemoaned a landscape full of 'loads of rubbish', 'neglected buildings' and 'weeds everywhere'; one spoke of 'constantly seeing blocked gullies'. 'I'm not sure Macclesfield is any better, or worse, than any other small town', one male resident reflected. 'But we do have a hideous litter problem. I live on a reasonably busy road into town, and on that road it's nearly always stuff that people throw out of their cars. I've actually seen people in traffic jams just empty their ashtray out of their window. It's disgusting.' For some of our participants, the 'throw away', littering culture that degraded the environment was something they found particularly distressing:

> It may be a small thing for most people, but litter distresses me. Our culture of throwing things out of car windows and having huge amounts of litter that doesn't biodegrade along the motorways and roadways, that concerns me a great deal. I'm not so bothered about the council not cutting the grass because the wildlife love it. But I am concerned about litter. I don't like the throw-away culture. (Female resident)

One common civic response to these felt concerns is litter-picking. In June 2022, there were over 140 routinely picked routes across Macclesfield – an indication that attending to the local environment is regarded as some kind of collective responsibility. They ranged from litter-picks organized on a regular basis by eco-conscious groups and businesses, to organized area-based litter-picks, to ad hoc groups advertised on Facebook. During walking interviews and participant observation with litter-picking groups in the town, those removing litter often reflected on its potential for environmental harm. To them litter was much more than an 'incivility', as this fieldnote attests:

> The litter-pickers paid particular attention to the area around the river, picking up pieces of plastic. 'There is always stuff like this on the banks and it will end up in the river and pollute the water'. One litter-picker made it their mission to scour the banks of the river and lean as far as she could into the riverbank to capture any plastic litter that could fall in and pollute. They knew where the usual places for litter were and moved beyond the

immediately visible litter to look under shrubs and hidden by vegetation spots. The litter-pickers' bags got fuller, and as the morning got on, I got used to seeing litter everywhere. The litter-pickers reflected on why people would drop stuff like this when there were bins nearby. There was no judgement, just puzzlement, and shaking of heads. The concerns were about what happens to the rubbish, the pollution it can cause, the escaping of plastics into a nearby river and the harm it causes to wildlife: 'Once it falls in the river it will be polluting for ever.'

The detritus of development

Potholes – and the generally frayed condition of the road infrastructure – were also treated as features of this degraded landscape. The presence and scale of potholes were a common reference point in conversations about the state of the town. On both neighbourhood and dedicated transport Facebook pages (such as Weather/Road/Traffic Problems Macclesfield) members shared photos of potholes alongside comments on repairs and their frustration at decaying infrastructure and a state unable to repair it.[1] The worry that 'no town I've come across has got potholes as bad as us', as one male resident put it, is not simply an aesthetic concern. People were anxious about the damage potholed roads cause to vehicles and the acute dangers they pose to the safety of cyclists. But potholes also send troubling signals about the condition and future of the town. They are a banal material meeting point between heavier cars, enduring austerity and ecological mutation. They bring people into tense relation with the outlooks, capacities and competence of authorities tasked with maintaining the roads (Annavarapu, 2025). 'The council still doesn't appear to be doing anything about it', one male resident complained. 'I don't know what is to blame for that non-responsiveness or that lethargy. I don't know if they genuinely don't care or that they don't have the resources. But it's the general sense of non-responsiveness and temporary solutions that frustrate me the most about the governance.' People relatedly worry that if the local state cannot maintain the roads, how can it be expected to address more serious problems or deliver on any more ambitious projects:

> We're slowly falling apart at the seams. You look at the state of the roads, they're disintegrating, certainly from what I've heard about what's going on in Cheshire East Council. We have a massive shortfall on just repairing roads. And I think that is the most obvious manifestation of what's going on. The libraries have been run down. We're just not looking after this country. (Male resident)

Given this, exasperation about the road infrastructure often spilled over into the concerns people express about what one might call the detritus of development. New housing development was a contested issue in the town – most obviously in respect of a proposed development on Dane Moss peatland, but also more generally.[2] But what kinds of concern does such development prompt? And can public concerns be interpreted as worries about disorder?

One typical feature of 'incivilities' is their sensory quality – one sees, smells or hears disorder (McClanahan and South, 2020). As Persak (2021: 52) puts it: 'visibility or exposure' is a 'precondition for experiencing offendedness'. She cites the example of particle pollution, which, she argues, tends not to provoke the same 'moral panic, security concerns and indignation [as] rubbish or beggars on the street or a person occupying two seats or talking loudly on the train'. Here, harm has to be visible – or at least sensed – to cause offence. How then does invisible harm (pollution, destruction of more-than-human habitats), signalling different threats (to oneself/family, one's town, the earth), on varying temporal scales (now, or in the future), feature in everyday contests over urban development, or in respect of what some are clearly minded to code as state incivilities? Unlike litter or potholes, particle pollution is invisible. But the increase of traffic and fumes from cars, and the planning action or inaction that cause them, are sources of daily offence.

In August 2022, traffic lights were installed at a roundabout to accommodate the building of a contested development to the west of Macclesfield. Broken Cross is one of three locations in the town designated as Air Quality Management Areas.[3] A large new housing development on 'green belt' land which would require the removal of peat had recently been approved, despite over 700 local objections. The installation of the traffic lights intensified local discussion of the development. One resident captured this in an open letter to their local councillor posted to the neighbourhood Facebook group. They bemoaned the 'chaotic' and 'dangerous' traffic control measures and 'constant stream of traffic and fumes blowing into our homes', before recounting the chaotic and disorderly daily trip to the local supermarket. What should have been an unremarkable passage to buy groceries had become an 'assault course', dangerous for the elderly, disabled people and pushchairs, 'causing harm and stress' to everyday users of the area.

The submitted objections to the development raised issues around air quality, traffic management, risk to pedestrians and the destruction of peat. Residents objected that 'you can smell and taste the pollution'. They described congested traffic as routine occurrence and reported recurrent experiences of flooding. Objections meshed concerns about harm to humans (road safety concerns and impact on human physical and mental health) with the harm to the environment that the proposed development would cause:

> It also brings great concerns about POLLUTION which leads to great worries about the health risks to my family with increased nitrate oxides, which increases the risk of cancer, heart problems, type 2 diabetes and asthma. This increase in houses has been added

> without any increase in the infrastructure. I am also very concerned about the removal of peat from the field. This is of great importance and great thought should be considered before removing any of it. We have seen in recent times the devastation caused by flooding. This area is flooded most of the year.[4]

An open letter to the (then) Conservative Macclesfield MP and local councillors by a group of residents captures the issues raised in objections to the scheme:

> We ask Mr Rutley to kindly raise our concerns with The Transport Secretary, Mr Shapps. Not altering Broken Cross will lead to increased pollution in an air quality management area with already illegal levels, potentially damage the economic viability of the town centre, and MOST IMPORTANTLY create immeasurable damage to local resident's respiratory systems, breathing in the pollution as they visit shops and pubs in the junction area, along with children walking to distant schools as there are no places in the vicinity of the development. This will lead to premature deaths, none of us want to see any young people in our area die in the way that Ella Kissi-Debrah did in London. (Macclesfield Nub News, 28 October 2021)[5]

The decision to allow the development was met with suspicion about the motives behind, and probity of, the process, a recurrent theme in local protests surrounding the development. 'We believe that this application is being driven by a mercenary financial motivation', one resident complained. 'This can only make the lives of all the local residents and wildlife and the environment poorer as a result of it being permitted.' Some of those objecting referenced the legacy such careless development would bequeath the town and their children, and situated the issue in the climate emergency:

> We are seeing the impacts of climate change now – so much rain, flooding and water to cause more flooding! We need to help our planet, not destroy its natural defences (trees, carbon) for greed and economic growth. This land is not suitable for development, it is an area of natural beauty, it maintains much wildlife, trees and peat which are all vital for our planet. Additionally local services are already overly subscribed, GP services, hospitals and schools. … Cheshire East Council, please; we as adults needs to leave our children a planet. What good are houses if the planet is destroyed?[6]

What we have observed in these stated objections are efforts to give form, and voice, to slowly evolving, often invisible, harms (cf Davies, 2022). This does not entail a blanket refusal to accommodate new housing (our survey and qualitative interviews in fact showed strong support for more affordable housing). But it does reveal a lack of trust in the willingness or capacity of authorities to make proper infrastructural provision for the town's growth or to reassure residents that environmental concerns and their attendant dystopian futures are taken seriously. This is echoed in the worries people register about housing developments that go ahead without attention being paid to the condition of the roads, traffic congestion, the provision of additional schools or adequate healthcare, or the impact on wildlife and the physical landscape. In the absence of care for such matters, development gets coded as another form of environmental harm. As one resident put it: 'We need to breathe in this town.'

Recasting everyday disorder

Revisiting Macclesfield after 25 years, we are conscious of continuities but also, more interestingly, of change. It would be tempting to interpret some of our participants' remarks about the things that trouble them locally – drugs, litter, local decline – within the staple frames of criminological and associated political discourses on anti-social behaviour and fear of crime, and to treat them as echoes of the 'culture of control' that typified the 1990s (Garland, 2001). Critical responses to such 'law-and-order' framings often attribute to those who voice concerns about crime and disorder, perhaps especially when it involves demands for the control of young people, strangers or other problematic figures, a degree of excess, or a tendency to uncritically accept a dominant ideology. In those cases, people are depicted as viewing their own experience through a lens imposed by media discourse, or political rhetoric, and as reproducing prevailing common sense on the matter. The result is a somewhat predictable inventory of issues that leads to a similarly unsurprising set of diagnoses and preferred reactions. Here, everyday security gets framed in a 'major' key.

To focus on such framings would, however, largely miss the point and fail to attend closely to important aspects of our participants' views. No doubt many of them would support firm action being taken to discourage drugs and other forms of anti-social behaviour. They want the town, or their part of it, to be defended against incursion using the traditional arsenal of police and criminal justice. But this framing, on the basis of our evidence, risks misunderstanding both the range and substance of people's expressed concerns and the attendant mood. That mood is one which takes anti-social behaviour to be less a harbinger of property or violent crime, as suggested in 'Broken windows'-influenced responses to social disorder (Wilson and Kelling, 1982), than a sign of environmental degradation – a 'proliferation of supposedly minor disturbances of everyday life' (Marres, 2026: 7) that have an incremental, damaging effect on the felt quality of the streets and neighbourhoods in which people seek to build and sustain passably secure lives.

If this is right, it underscores the value and importance of recasting everyday security in a minor key. It intimates a need to lift questions of social ordering out of a narrow crime and punishment framing, to speak not of 'incivilities' but, perhaps, of 'in*civic*ities', and to relocate concerns about chronic harm in a politics of the 'good city' (Amin, 2006; Kern, 2020; Lundberg et al, 2024); or debates about 'social infrastructure' (Klinenberg, 2018), or questions to do with the ecological governance of place. It is to these matters that we turn in the next and subsequent chapters.

FOUR

Erosion of Public Things

During an interview with one long-standing resident, who had made clear her view that Macclesfield was 'very friendly, a very nice place to live', we asked what, if anything, she disliked about the town. 'There are bits that feel a bit grubby', she observed. 'There's quite a lot of derelict buildings which I think could be used for housing. There are lots of sites around the centre which should be developed or should be used. But they're just sort of left abandoned.' She continued:

> They look abandoned, so they make it feel as if the town is abandoned and they look derelict. And plants start growing, and then the plants catch bags in them or weeds, and nobody tidies it, and nobody cleans it, so it just looks abandoned and uncared for. So, it gives the impression that it's not a cared for community, which generally is incorrect, but it feels that way.

In *Palaces for the People*, Eric Klinenberg identifies a path not taken by urban public policy since the 1980s (Klinenberg, 2018: 60–3). He recalls that, in their globally influential theory of 'broken windows', Wilson and Kelling (1982) identify a number of disorderly behaviours and objects that, left unchecked, can send neighbourhoods into decline and crime spiralling. 'A piece of property is abandoned, weeds grow up, a window is smashed', Wilson and Kelling write, before listing the range of disorders that follow: panhandlers, groups of untended youth, the public sale and consumption of drugs, vandalism, graffiti, and so on. What followed, we now know, was aggressive street-level policing of low-level disorderly behaviour: harassment, stop-checks, arrests, especially of minority youths (Harcourt, 2001).

But what, Klinenberg asks, would have happened if, instead, city authorities had put equivalent effort and resources into maintaining public safety, and staving off neighbourhood decline, by 'cracking down' on neglected or abandoned buildings. This could have meant moving fast to fill them with temporary, pop-up projects, acts of what architect and planner Jaime Lerner calls 'urban acupuncture'. 'The quicker an abandoned lot is occupied, the better', Lerner writes, 'and preferably with something even more attractive or lively than before' (Lerner, 2014: 38). It could also have meant restoring abandoned buildings as affordable homes or using them to facilitate what Klinenberg calls 'social infrastructure' – libraries, cafes, cinemas, daycare centres, sports clubs, small businesses of various kinds. Public policy towards urban disorder might have looked and felt markedly different.

This chapter is about infrastructure, both physical and social. In the first half, we report on the local concerns expressed about the condition and possible futures of the town. We focus mainly on aspects of the town centre but touch on related concerns about such things as roads, pavements and public transport, and parks and public spaces. This is a tale about enduring austerity and its felt local effects; it is also a story of decay and loss, not only of vital facilities, but of democratic agency. We then reflect on why such facilities matter to people and tease out what is at stake when people assemble around them. Here, we pick up and develop Klinenberg's argument about the value of social infrastructure. We suggest that these infrastructures matter because of their status as 'public things' (Honig, 2017) – things that people experience, share and contest in common, things that embody what it means to make a passably satisfying and sustainable life in this place.

Decaying infrastructures

Macclesfield Town Hall stands at the heart of the town's central area overlooking the Market Place. It is a grand civic building dating from 1823–4, designed by Francis Goodwin in the Greek Revival style. One wouldn't describe the town hall as abandoned property. It houses council meetings and offices, and its function rooms can be hired for gatherings and events. Yet there were those in the town who told us that it feels under-utilized, its promise overlooked and unfulfilled. For most of the time the town hall's doors were indeed closed and its downstairs rooms seemingly empty: it hides its public face. To some among our participants, this appeared symbolic of the relation that local political authority has come to assume to the publics it serves: remote, shut-off, unreachable. It felt like an opportunity has been missed to connect the building to the town by opening it up to community groups and uses, thereby placing it at the material heart of the town's public life. As one active local campaigner put it: 'Even when it is being used, the doors are always shut. It has the capacity to be incredibly transformative, to drive a sort of recovery and reinvigoration of the town centre, to offer stuff beyond retail. But the willingness to open dialogue is hugely limited.' There are many reasons – both logistical and financial – why these possibilities have not been realized. But the sense lingers of the civic imagination having been dulled.

Concerns about local infrastructure – and what its state of decay and disrepair might indicate about Macclesfield's prospects – crystallize around several objects; but foremost among them is the condition and future of the town centre. The town centre has been a topic of local concern and emotive debate for some considerable time, as well as being subject to various plans for regeneration stretching back for over a decade. Each, in turn, has faltered. Not surprisingly, it featured prominently in our research. Much of the attendant concern focused on what was variously described as the 'run down', 'quiet', 'tired', 'dreary' and 'really, really sad' state of the town centre, and the sense that relevant governing authorities were guilty of 'institutional negligence' (Murphy, 2012: 9). The following exchange is indicative:

> Interviewer: If you were asked by an outsider who knew nothing about the town, how would you describe it to them, the town today?
>
> Male resident: It's a town of two halves really. On the surface it can look quite nice. It's hilly. It's surrounded by beautiful countryside, the cobbled streets, quite a few old buildings there. But if you scrape beneath the surface, you see there's an awful lot of decline. I think it just looks really sad. It looks really sad having lots of empty shops. It creates a sense of dereliction. So that's the prevailing atmosphere.

Worries about the felt decline of the town centre took several forms. Some coalesced around the presence of – troubled and troubling – people. The closure of unviable shops has left the town centre replete with vacant buildings, some of which now offer informal shelter to those without stable accommodation. 'The biggest thing that bothers me about Macclesfield's main street is the beggars that live in the shopfronts', one woman reported. 'We've got about three of those who've been camping and there's a whole paraphernalia of tents and rubbish, and maybe a shopping trolley. I find that really off-putting.' Another participant remarked of an abandoned building on the edge of the town centre: 'when it was empty, loads of kids started going in it and there were drug users in there with needles'. A police officer we observed described how had he stumbled upon a 'cannabis factory' in one such abandoned property.

The attendant worry is 'that litter, along with some of the shops closing, gives people the idea that this is going to fall into one of these ghost towns that's just going to be full of drug addicts and charity shops' (female resident). One would expect this to evoke a certain amount of anxiety, and among some participants it clearly does: those viewed as homeless are seen by many as unwelcome, a blight on the town. But fear or distaste were by no means the prevailing emotions. There was, perhaps more prominently, an evident sadness that 'our' town has become a repository of visible human misery.

A related set of worries focus not on the presence of people, but their absence. Residents commonly spoke of the town centre in ways that referenced its 'emptiness', or 'ghostly' feel. 'It's just very quiet in general, which is a bit weird', one woman reflected. Another – channelling Jane Jacobs (1961) – described the empty town centre as 'unnerving', lacking 'the reassurance' provided by 'busyness'. One puzzle here is why a town of Macclesfield's size cannot support an economically and socially vibrant town centre. A related concern is whether the town's middle-class residents typically look outwards to the prosperous Cheshire towns between Macclesfield and Manchester for their leisure activities and have evacuated the town itself – a trend confirmed in several of our interviews. As one male resident observed: 'It's a very middle-class town, I think. An awful lot of commuters never come into the town centre. We have a population of 50,000. Looking at this high street, you wouldn't think there was anyone shopping here. Lots of empty shops.'[1] The common refrain is that the town needs people, a refrain commonly voiced by people conscious that they are co-authors of the problem that so concerns them: 'Perhaps I should make an effort to go up there, you know. Because that's what it needs, it needs people, ordinary folks, just to go up there, and perhaps it will feel a bit better. It's a double-edged sword isn't it. It's difficult' (male resident). It is, as another resident put it, 'a sort of vicious circle'.

During our research, these worries were brought to a head by the closure (and partial relocation to an out-of-town shopping development) of Marks & Spencer – the last major retail outlet in the centre of Macclesfield. One local professional reflected on this:

> Marks & Spencer is shut now, and it just looks really rubbish. I don't know if people can make things happen, but going up Mill Lane and walking up towards the town hall is getting more and more depressing and more empty. And why would you go there if there's that sense that this is all getting a bit fractured socially? So, that's what I'm really, really worried about.

The question of what has happened to Macclesfield town centre – and what to do about it – is clearly a local instance of a national story (though people would often remark that other towns in the region appeared to be faring better).[2] Town centres across the UK are facing considerable challenges in the face of entrenched austerity and changing patterns of consumption and leisure (Kelsey and Kenny, 2021). This though *is* the main challenge that residents wanted their local authorities to tackle: our 2022 survey found that nearly one-third of participants placed 'regenerating the town centre' as their top priority for action (see Table 4.1). There is also ongoing (local) debate about what might be done to spark regeneration. Some call for changes to planning laws, or reduced business rates, or bemoan the cost of parking. Others call for pedestrianization and related measures to curb car traffic, proffer ideas for active place-making, or propose revitalization of the town centre through high-density affordable housing. In the face of this challenge – how to regenerate and sustain the town centre in ways that cater to the aspirations, needs and budgets of all ages and socio-economic groups in Macclesfield – some among participants voiced their own fears and hopes. 'It really does need planners with imagination', one remarked.

Table 4.1: Local government priorities: what would you say are the top three things that the local councils (Cheshire East and/or Macclesfield Town Council) should be doing in your area?

	2022 (%)
Regenerating the town centre and the shopping area of town	32
Improving the roads	22
Tackling anti-social behaviour	16
Making homes available to people who need them	12
Looking after/improving public spaces	5
Providing more facilities for residents	4
Improving public transport, walking and cycling facilities	3
Improving accessibility around town	3
Educating people about online safety	1
Dealing with rogue traders	<0.05
Other (specify)	1
Don't know	<0.05

Note: Weighted total vote. Percentages based on weighted total vote where first choices score 3, second choices score 2 and third choices score 1.

'Particularly as town centres right across the land are going through this critical moment, where big chain stores are moving out because of online demand. It gives opportunity, possibly.' Another male resident voiced the following ambition for a change of direction: 'From my own perspective, it's really about a town which is easy to travel around on foot, or by bicycle. And where nobody's frightened of using our streets, and a town where streets are caring communities, rather than rat runs with speeding traffic.'

So the slow decay of the town centre troubles people, here as in many other towns across the UK, because that decay is read as 'a sign of its approaching death' (Hage,

2021: 4) But the town centre is also a 'condensing symbol' (Turner, 1970) for diffuse concerns about the condition of Macclesfield's social and physical infrastructure, which was widely apprehended to be crumbling and in need of repair or investment. The second highest priority for action in our 2022 survey was 'improving the roads'. The potholed condition of local roads is 'believe it or not, the main problem for people round here', one resident exclaimed; we have described these concerns and their meanings in Chapter Three. But people also raised worries about broken or cracked paving on local footways, with all the danger that brings for pedestrians, as well as the unwanted incursion of nature into local infrastructure in the form of overgrown shrubbery and untended weeds. A further concern surrounded the 'awful' state of public transport within the town (the connections to places beyond Macclesfield are relatively good), especially from the outer housing estates to the centre, such that 'if you wanted to go out in the evening, you are really dependent on a car or taxi' (see further, Loader et al, 2025). The following exchange encapsulates this range of concerns:

Interviewer:	Is there anything that troubles you about Macclesfield, or the immediate neighbourhood in which you live?
Male resident:	Nothing jumps immediately to mind. But I will admit, it does require a certain level of infrastructure investment, to say the least. I've already mentioned street-lighting. Obviously, potholes in roads, road services certainly are pretty atrocious. I think if you don't keep up to a certain level of infrastructure and you let things deteriorate, there's going to be less of a reason for people who do want to live in the area to … it comes to a point where it's off-putting.

We can here return, briefly, to the matter of drug-detritus, littering and associated forms of disorder described in the previous chapter. 'Tackling anti-social behaviour' was the third highest priority expressed by residents when asked in our survey what issues they wanted local authorities to focus on (see Table 4.1). Part of what animates concerns about disorder is that it sometimes occurs *in* and involves the spoiling *of* public parks. Macclesfield's parks are objects of immense social value and deep affinity. They are nostalgically wrapped up with childhood memories. They are green spaces that break up dense urban landscapes. They are places of escape, rest, physical exercise, association and chance encounter. They are sites of children's play free of the risks of road danger (see further Churchill et al, 2017). For these reasons residents are highly attuned to danger in local parks (kids drinking alcohol, signs of drug-use, noises emanating from the dark), dangers that are gendered in significant ways (Barker and Holmes, 2021). And they are attentive to damage being inflicted on parks and the various facilities they provide, whether in the form of litter, graffiti or vandalism.

Local contests over parks are frequently funnelled through this relation of affinity and danger. In our discussions, residents often shared stories about 'beer cans', 'needles' or occasional acts of violence, and exchanged views on whether teenagers congregating in parks were either 'kids being kids', or else threatening others and placing themselves at risk. Something similar was at play in local reaction to a story in the *Macclesfield Express*, in April 2022, headlined 'Family park "overrun by gangs and drugs"'. Local residents refused to be taken in, still less frightened, and sprang to the park's defence. 'Bad and unsavoury things happen in this park, as everywhere', they conceded in a letter to the paper, before bemoaning the 'unhelpful scaremongering' and reiterating the park's rich and varied social value: the regular park run, allotments, the cafe, skate park, and so on. Parks are social infrastructure that people care about. They want them looked after (cf Smith et al, 2025).

The same is true of roads. Less than half a mile away from Macclesfield Town Hall, in 2018, on a residential street in a neighbourhood adjacent to the town centre, a large sinkhole appeared, the tarmac reclaimed by the earth. Over two years on, the hole remained unrepaired, despite numerous attempts by the authorities to do so: the council called it 'an extremely complex piece of work'. The prolonged road closure had several effects. Residents marvelled about the 'return' of more-than-human life to their street. 'Nature has re-claimed the road', one reported. 'We've got buddleia in there and tomatoes. Among the weeds we've got a couple of trees growing there, there's definitely a maple.' But the hole was at the same time a source of anxiety. 'It's also a rat sanctuary. … There's an open sewer; rats can go up and down into our toilet. I daren't put the seat down on my toilet in case I have rats in the bathroom that's my fear.' Residents further worried about potential gas leaks, and about vulnerable people being cut off from emergency assistance: ambulances and the fire service could no longer access the street.[3] This experience – and the absence of timely or effective repair to basic infrastructure – galvanized them. They sent emails, wrote letters, organized a petition. When none of this worked, exasperated residents brought the hole to the world's attention via a music video posted on social media. The video – 'Macclesfield Sink Hole Day 700' – has been viewed 9,500 times on YouTube.[4]

On the value of public things

The Weston estate lies a couple of miles west of Macclesfield town centre. Originally built following the Second World War as council housing, it is now a mix of houses purchased by their former tenants and social housing. It stands among Macclesfield's most disadvantaged neighbourhoods. Problems of crime, drug-taking and kids hanging around feature prominently among the concerns of local residents, as does the physical condition of the area. So too do problems of addiction, loneliness, vulnerability and potential exploitation. During a walking interview through the Weston estate we conducted with one long-standing Macclesfield resident, the woman surveyed the streetscape around us. She paused, before noting:

> If you come outside of your front door, it doesn't matter how good your house is or how nice your house is. If you come outside your front door and it's full of filth, or its just endless concrete and tarmac, it's not very conducive to coming out and talking to the neighbours, perhaps having a cup of coffee with them, or setting up a little cafe or something. I think we've got to treat the town as our home and not just our house as our home.

It is easy to dismiss, even mock, the everyday concerns described in this chapter and the last. They can seem minor and trivial. When set against the problems of vulnerability and violence in domestic settings that increasingly preoccupy the local police, they can seem misplaced, even ridiculous. Senior police officers we spoke to often dismissed them as such, troubles unworthy of police prioritization, even attention, in the face of the more serious problems with which officers have to deal. These public troubles can also seem backward-looking, parochial and conservative, components of a nostalgic sensibility that seeks to resurrect forms of communal life – and harmony – inescapably washed away by economic, cultural and technological change. Macclesfield's 'whiteness' can fuel these suspicions. People are yearning for a social world that has ceased to exist.

Such dismissals are too hasty. They are also a mistake, resting on a partial reading of these concerns. As we noted in Chapter Three, worries about a decaying public realm and its consequences can take shape as demands for action against disorderly or troubling *people* and calls for some kind of police intervention, narrowly understood. It would be remiss to pretend that such framings are not an aspect of public affects and discourse. But for the greater part, the concerns that matter to residents, and their attendant demands, are not focused on the policing of people; rather, they are about unattended, neglected or abandoned *places*. These situated judgements can be said, in this regard, to be inchoate versions of the critique of public policies on urban disorder outlined – in a US context – by Eric Klinenberg in *Palaces for the People*:

> To this day, however, most policies aimed at reducing crime focus on punishing people rather than improving places. The President has called for a national 'stop and frisk' program; the attorney general wants more severe sentencing; advocates of 'law and order' are resurgent. We invest little in housing and far less in safe sidewalks and neighborhood amenities like libraries, senior centers, and community gardens, which draw people into the public realm and put more eyes on the street. (Klinenberg, 2018: 59)

It is also evident that people's worries about decay and disorder are not disaggregated into separate problems, nor neatly parcelled up into those that involve criminal infraction and those that do not. All are forms of environmental harm. The detritus of drugs, litter, cracked pavements and potholed roads, awful public transport, vandalism, nuisance youths in parks, a declining town centre: these each entail decay, damage and loss to elements of an ecosystem within which people strive to make and sustain passably viable lives. Something of value is being eroded or lost. But why are these objects of concern? What, exactly, is being lost?

These assorted problems matter to people because they entail the degrading of what Bonnie Honig (2017) calls 'public things'. Physical infrastructure such as roads, pavements, street lighting, sewage systems, and public transportation and social infrastructure of the kind identified by Klinenberg, matter because they 'shape the way people interact' (Klinenberg,

2018: 5): they enable or discourage social interaction and foster or impede mutual support. 'Democracy is rooted', Honig (2017: 4) argues, 'in common love for, antipathy to, and contestation of public things'. 'Without public things, we have nothing or not much to deliberate about, constellate around, or agonistically contest' (Honig, 2017: 5). 'Public things are part of the "holding environment" of democratic citizenship; they furnish the world of democratic life' (Honig, 2017: 5). 'When infrastructure crumbles, it is not only infrastructure that crumbles, but also its constitutive gifts of stabilization, integration and adhesion' (Honig, 2017: 6).

Social and physical infrastructure is, in these respects, a vital mediator of belonging, and of the sense of everyday security supplied by that belonging (Loader and Walker, 2007: ch 6). It establishes a series of frameworks that enable engagement with others and facilitate support from others. It is integral to people's sense that the place they inhabit is not simply a collection of houses for 'individuals and their families' (to coin Margaret Thatcher's preferred alternative to 'society'[5]) but, instead, is composed of things that offer those individuals and families some kind of meaningful communal life, give the place they inhabit some kind of vital publicness. As one woman observed of the town's library:

> I love being in the library and the librarians … I don't know any of them. But it's little interactions that you have with people that make you feel part of a community. I think more so than the big things. I think Macclesfield has the opportunity for that kind of thing, the small interactions, as well as the big ones.

Those objects also provide 'an intangible bond that forges a sense of belonging' (Lerner, 2014: 10) by embodying a continuity between past, present and future that sustains collective memories and identities. Attachments to these things 'allow people to inhabit and make liveable worlds, especially in damaging conditions' (Anderson, 2022: 393). No wonder they express dismay at their state of disrepair and possible loss:

> My son, he's 29, and he smirks at me sometimes because I do harp back to a golden age and he spends more of his life online. But I think town centres are important. People like to eat out, they like to go for a drink, they like to go to the cinema. You don't always have to do that at a town centre, but I think it contributes something to the social life of a place if it's got a centre where people can congregate. And I do worry that Macclesfield town centre will continue to decline and have fewer and fewer shops, and that's bad enough. But it just looks really odd, like a ghost town. (Male resident)

During one of our interviews, a young woman was reflecting upon how the town centre was changing, when she remarked: 'Almost every time I walk through the town something suddenly wouldn't be there and the storefronts are left.' This chance observation – '*suddenly* something wouldn't be there' – offers a small but telling reminder of the ways in which the future of social infrastructures is not easily or sensibly separated from questions of democratic politics. The coming and going of shops, the shifting location of the post office, and the occupancy or abandonment of public buildings, can be – indeed often is – put down to the operation of the market economy. But to local people it can be experienced quite otherwise. These sudden, unpredictable mutations offer 'glimpses of a public culture shifting over time' (Stewart, 2007: 26) in ways that signal that something is not quite right. They are also an index of people's lack of agency over the place in which they live and over matters to do with the quality of *their* street, *their* neighbourhood, *their* town centre, *their* lives. It is a sign of a world that exists largely beyond their control, a world in which they appear to have little choice but to accept the 'austerity-driven erosion of the public sphere' (Stears, 2021: 189).

The value of public things doesn't inhere, in this respect, in some magical capacity to generate consensus or efface conflict. Nor should we want or expect them to do so, as Berlant (2016) warns in their critique of complacent and overly cosy appeals to the idea of 'the commons'. Public things help constitute publics because they are things that 'we' care about, look after, and use or share in common: 'they press us into relations with others' (Honig, 2017: 6). But they also constitute democratic publics by providing things for 'us' to assemble around, contest and argue about. They form part of the circuit of democratic life that we act upon, and that can act upon us in ways that both bind and unbind (Honig, 2017: 7). There can, as Honig nicely puts it, be no 'subjects without objects' (Honig, 2017: 5). Public things are sites for the cultivation and exercise of democratic agency. Their erosion is felt as a diminution of democracy.

FIVE

Driving Disorder

We did not return to Macclesfield intending to ask people about cars, nor expecting local people to complain to us so frequently about the impact of motorized traffic on their daily lives. In our earlier study, a quarter of a century earlier, cars had featured in various ways. Damage to, and theft of, vehicles was a prominent concern in the town at that time, as it was nationally. It stood high among crime prevention priorities. Car-owners became often viscerally angry about the attention they received from the police as speeding or (suspected) drunk drivers. They resented being targets of an institution whose purpose, in their view, was to protect 'us' from 'others', and felt the police should spend their time 'catching real criminals'. Cars were implicated in the threat posed by 'travelling crime' – Macclesfield's good road connections enabling offending outsiders to access the town from nearby Manchester and Liverpool. We also encountered some grumbling about the state of Macclesfield's congested road system, though our comments upon this were limited to a single footnote:

> So too has Macclesfield come to be dominated by that other pervasive object of contemporary urban living – the car; something that was registered in the prominent place traffic congestion was accorded when people were asked to select the 'worst thing' about Macclesfield. One notable sign of this was our respondent's fondness for remarking that Macclesfield was the only place they knew where 'they' (the Council) had 'put a by-pass through the

> middle of the town'; a tale whose resonance is indicated by the fact that it was told to us in relation to no less than three different roads. (Girling et al, 2000: 185)

For the most part, however, people spoke of traffic problems as they did about the weather; a fact of (British) life that had to be endured, but could not be changed; an item of small talk to break the ice before an interview or focus group commenced, only to be dropped in favour of 'serious' matters once the tape recorder was switched on.

In December 2019, at the start of our fieldwork, we got an early indication that local sensibilities might have changed when, at a public meeting in Macclesfield Town Hall, called by the Police and Crime Commissioner and Chief Constable of Cheshire, concerns about parking at school drop-off were raised vociferously by local residents. Similar concerns about the impact of cars on the social ecology of the town then resonated through our enquiry. The planning applications we analysed in 2021 were full of objections relating to traffic congestion, parking and air pollution. The volume, speed and parking of cars is a source of regular discussion on local social media, especially neighbourhood Facebook groups, and in response to police posts. The car featured prominently in discussions about the condition and future of the town centre, and when asked by us what filled their 'post-bag' from constituents, town councillors replied without hesitation: 'traffic'. The concern that most often finds expression on the police Residents' Voice platform was speeding and demands for greater enforcement. 'Badly parked cars' were high among concerns about anti-social behaviour in our local surveys. Conflict between parents and residents prompted by children being driven to and from school was an issue across many Macclesfield schools. Our observations revealed that the regulation of these conflicts was a recurring task of neighbourhood police. Cars, it seems, were no longer being treated like the weather (see, further, Loader et al, 2025).[1]

In this chapter, we report on the main lines of concern and contestation that surfaced around that most mundane object of everyday life: the car. We then seek to re-describe these emergent sensibilities in ways that tease out how automobility has come to be treated, and may fruitfully be theorized and investigated, as central to questions of urban order and disorder. 'Never make transportation an issue by itself', Andre Gorz (1973) once advised. 'Always connect it to the problem of the city.'

Forms of automotive disorder

One winter morning during our fieldwork, the patrol sergeant whose shift we were observing received a radio message about a traffic collision on the outskirts of the town centre. He decided to respond. When we arrived at the scene, two officers were already in attendance. An elderly male driver had crashed into a parked car. The officers were concerned that that man was not fit to drive, though the driver insisted the problem had to do with his new spectacles. They asked him to take an impromptu eye-test, by reading a shop sign about 20 metres away on the opposite side of the road. The driver could not make out the words. The officers told him he should voluntarily surrender his driving licence and that they would report the failed eye-sight test to the national licensing authority. They would take the driver and his car home. It was a poignant moment. We had witnessed the abrupt end of a man's long motoring life, and with it, no doubt, his sense of autonomy (Róin et al, 2025). But this man had been driving a two-tonne machine around Macclesfield, when he could barely see. It was a reminder of the serious environmental risk created by this assemblage of human and object; the police had disassembled a dangerous hybrid actant (Loader, 2026).

Our discussions on the matter of cars were situated both in time and place. The research was conducted *before* urban mobility in the UK transmogrified into a national 'culture war' issue in 2023, sparked by outrage on the political right about low traffic neighbourhoods, ultra-low emission zones, congestion-charging and 15-minute-cities, measures the Reform Party leader Nigel Farage portentously dismissed as 'climate lockdowns' (Marquet et al, 2024). Our research was also conducted in a town where little serious or sustained action was being taken by local authorities to restrict car use, encourage public transport, re-allocate space to enable safe active travel, or in any significant way alter the infrastructure that 'generate(s) the ambient environment of everyday life' (Larkin, 2013: 328). Macclesfield is, in this respect, like most English towns. Cheshire East Council has some aspirational policies that point in the direction of de-carbonizing transport, but that was about it. Cars remained 'unmarked' in local public policy (Zerubavel, 2018) – in part, no doubt, because authorities stand wary of disturbing those residents who assume a 'motonormative' disposition to the social world (Walker et al, 2023) and take driving for granted as 'what normal people do' (Paterson, 2007: 223).

Yet we also observed signs that cars *are* today being 'marked' by local people as a source of trouble and irritation, objects that impact negatively on their sense of Macclesfield as a safe and liveable place to inhabit. That disquiet took several forms. The most overarching local concern had to do with congestion: the sheer difficulty of moving around (by car, as well as by other means) created by the sheer volume of road traffic. 'It's a town which is not built for the motor car', one resident reflected, 'but there are far too many cars'. 'Don't try to get around in a car', another warned. 'There are certain times when it's especially bad, particularly school drop-off and pick-up times', he continued. 'It's just gridlocked. There's an awful lot of children that get taken to school in cars.'

Observations about congestion were typically made during conversations about how the town has changed over recent decades.[2] Some here drew contrasts with the distant past, conjuring up a time when public space was child- rather than car-centred: 'I used to live on [a main road], up by the Weston', one man recalled. 'We used to be able to play in all the streets around there. You would actually play football in the streets. You would cycle up and down the streets. There were cars around, but not in significant numbers.' For others, local experience of the first COVID-19 shutdown in 2020 brought the car question into sharp relief and briefly offered a glimpse of a different kind of place: 'It was lovely in that it was quiet, and all the traffic stopped.' Other residents projected forwards, focusing on the likely impact of new housing developments that bake-in future car dependency (New Economics Foundation, 2024). In respect of 'all these new houses, all these new cars', one female resident observed, 'there's the sort of micro-level of concerns about the increase in traffic, the increase in pollution generally, the impact on the environment … there's been quite a lot of construction'. All these anxieties were variants on the theme of a town suffering from 'urban cholesterol' caused by 'the buildup of excessive automobile use in our veins and arteries' (Lerner, 2014: 95). As one male town centre resident reflected: 'It's very car-centred. And the more you get interested in seeing how the volume of cars restricts people walking and cycling, and makes their lives more difficult, the more annoying it becomes. From my perspective, that's one of the town's chief problems.'

Traffic congestion is the backdrop against which people connected the car to questions of everyday security. Our two community surveys found that the 'signs of drug selling and use' were the biggest safety concerns expressed by Macclesfield residents, as we saw in Chapters Three and Four. But car-related disorders were not far behind (see Tables 3.1 and 3.2). Speeding and badly parked cars were next in line as prominent concerns when people were asked *both* about 'problems in your area' *and* the 'things that make you feel unsafe' in the town. In this context, speakers sometimes drew explicit comparison between automotive disorders and the seemingly more serious safety problems that typically feature in public discourse. 'Can

I say we haven't mentioned crime at all', one member of an environmental campaign group observed towards the end of our discussion. 'I don't know whether we were going to. But I don't feel, I don't feel it's a big issue, for me anyway. I feel safe in Macclesfield, except when I'm on my bicycle.' Reflecting on her daughters' safety, one mother used a similar hierarchy of danger to couch her concerns:

> Well, car safety's my biggest one. I mean, I don't think I worry about … in terms of them going out on their own, I'm not somebody who thinks about them being abducted by strangers. It's crossing roads. Because I think that's their actual biggest risk. So, I think again, there's lots of places, there's a lot of traffic and lots of cars and I worry about that for their safety. I think that's it really.

These mundane concerns about automotive harm are brought into sharp focus by one particularly troublesome local matter: the school run. Over a third of children in Macclesfield were routinely driven to school; only 1 per cent cycled. There were no 'School Streets' schemes in the town.[3] At the aforementioned public meeting held by the then Chief Constable and Police and Crime Commissioner in Macclesfield Town Hall in December 2019, complainants spoke of 'running the gauntlet' of speeding cars on the way to school, and of 'tempers flying' due to inconsiderate parking on the pavement around school. The school run was also named as a recurring problem in the town in several interviews and group discussions, 'traffic outside schools' being variously described as 'horrendous' (female resident, focus group) or 'mayhem' (council transport official). It came more fully to our attention during our observations with local police. Schools across Macclesfield were an ongoing matter for the police in this respect and sites of routine patrol by neighbourhood officers – as one police officer put it: 'It's chaos when I'm not here.' The following fieldnote illustrates some of the issues at stake:

> We went to patrol outside [X] Primary School. Being here, the PCSO said, had two purposes.[4] One was to be seen and to get to know people. Lots of the children and one or two parents said hello as they passed the gates into school. But the PCSO was mainly here to regulate parking and car drop-off issues. He explained the problem of the dangerous junction and those created by parents parking on the pavement blocking residences or dropping off in the no stopping zone outside school. The road is narrow and sits on a corner. The PCSO kept a watching brief and explained to me what his rules were: pavement parking is okay if a double buggy can get past; no parking in ways that are dangerous or obstructive; and no dropping off in the no stop zone. One driver just did this – he pulled up in an Audi SUV and stopped to let his child out. The PCSO approached the car and explained that he couldn't stop here on the yellow lines. The male driver barely spoke but had anger and irritation etched all over his face: as if he couldn't believe he were being reprimanded by the police for such innocuous everyday behaviour.

The school run is one key backdrop around which annoyance about speeding and parking crystallizes. Concerns about the former are especially acute on local residential streets, especially those which have become 'rat runs' for through traffic. In a local media report, a resident of one such street relayed her concerns, especially with respect to children's safety. Barbara said: 'I live on Bond Street and it's very concerning the speeds some people travel at. Especially as we have a primary school and lots of little side roads, often where you have to edge out slowly to see/hear oncoming vehicles due to the parked cars.' She then urged the local police to 'Keep up the good work!'. Another resident echoed the sentiment: 'Well done. Must be like shooting fish in a barrel on Bond Street. Crazy road to drive quickly on.' Deborah said: 'I also live in the Bond Street area. The speeds the people come down that road at is shocking.'[5]

Those walking to school are also confronted by obstructed pavements. In fact, the presence of cars is often most keenly felt, not when they are moving, but when they are static, or parked. Cars must be left somewhere. Most of the time cars in urban settings are not in motion but immobile – the average car in the UK is parked 96 per cent of the time.[6] Motionless cars actively perform work on the urban environment (Kurnicki, 2021). Cars not only require land for parking infrastructure – most obviously in the form of carparks, kerbsides or private driveways. Cars stored on the street temporarily take control of, and privatize, a portion of public space. When parked on the streets they collectively become part of the physical environment, in ways that have material effects on 'what, when and how people do things' in urban space (Kurnicki, 2021: 2). As Bloch (2021: 139) puts it: 'Struggles over parking are no small matter given the proportion of car space that dominates neighbourhoods.'

In Macclesfield, parking issues are most pronounced in the 'small, quite narrow' streets surrounding the town centre because, as one resident put it, 'there's just not the space is there. The roads weren't made for the number of cars or the current size of cars' (female resident, focus group). The basic problem is too many residents' cars – of increasing size and weight (Brand, 2024)[7] – chasing insufficient space, coupled with narrow streets that encourage pavement parking. The result, one young male participant put it, is that in streets 'built in 1850', before the advent of the car, 'people just get wound up at each other about parking'. Conflicts about 'bad' or 'anti-social' parking also surfaced across other neighbourhoods in the town. During our research a Facebook site – 'Dangerous Macc Parking' – appeared as a place for local people to report bad parking.[8] This is a local instance of what has become a growing online phenomenon. Variants of @badlyparked X, Facebook or BlueSky accounts can be found in many towns and cities in the UK and enable people to name and shame drivers and post photographic evidence. Posts to such sites commonly tag the relevant police force or local authority. There are also national X and BlueSky sites that perform the same role, such as @YPLAC.

Sometimes the concerns prompted by bad parking had to do with what it signalled about drivers' lack of care for the safety of others: 'I get there are more cars today than years ago', one resident posted on a local social media platform. 'But peoples drives are blocked, junctions are dangerously obstructed, pathways blocked. Nobody considers others safety. Not just here. It's going on everywhere.' Others complained about 'selfish people who park on pavements' with scant regard for other users of the occupied space. As a female resident from one of Macclesfield's estates put it:

> I've just had to walk on (X Road) with my baby in his pram because some idiot has left no room at all when parking!!!😒 I tried to get passed but the pram's wheels wouldn't fit. If someone in a wheelchair comes across this, they're going to have to climb a steep curb and walk in the road! So bloody selfish.

Against this general backdrop, one parent posted about 'the idiot that parked their car blocking an entire pavement for the past two days at school drop off'. 'Don't worry', she wrote, 'we will all walk our children in the road because you can't be bothered to walk a little bit further'.

Two prevalent themes recur through these complaints about car-related social disorder: that these problems are caused by a minority of drivers, and that this minority displays a lack of care or concern for others, or failure to respect the rights of those who walk to, or live near, schools. These drivers are imagined as exemplars of what Mitchell (2005: 96) calls an 'SUV model of citizenship', autonomous and insulated actors with little regard for the collective good: 'we ride high and sovereign; we are masters of space; we are safe against all who might intrude, all who might stand in our way'.

There are, however, several features of these everyday car troubles that undermine this act of sense-making. Much of the behaviour about which people worry and complain is legal – annoying yes, unsafe perhaps, insecurity-generating, it would seem so; but illegal, no. Drivers can drive 'too fast' at 30mph on residential streets with 30mph limits. People can park legally but in ways that show a lack of consideration for other street users and scant concern for the quality of neighbourhood life. Parents can make short car journeys to school that others judge to cause urban despoliation and environmental harm, but there is no law sanctioning this form of mundane car-use. But it is not simply that these activities are legal. It is also that the car problems about which people express concern flow from behaviour they also engage in, perhaps routinely. Haven't we all driven too fast, close-passed a cyclist, parked where we shouldn't, sat with engines idling, or driven our children to school? The majority, in other words, may be implicated as authors of their own peril. Let us explore this further.

Framings of automotive disorder

People's relationship to place, and to the authorities governing that place, is mediated by infrastructures. 'Slowly formed over time, infrastructures are made by and constitutive of diverse political rationalities, past and present' (Anand, 2017: 12). In the case of transportation, infrastructures create material and social affordances for certain kinds of movement (whether that be walking, cycling, driving or taking public transport), while rendering other mobilities more difficult, even impossible. They also configure in particular ways the relation between moving through, and dwelling in, place (Sennett, 2018: 35–7).

There is an important infrastructural context to the forms of car-related chronic harm we encountered in Macclesfield. One element of that context concerns cuts to public transport in the town which in the view of many residents have made the local bus service at best patchy and on Sundays and in the evening close to non-existent. The second is the limited scope of dedicated cycle infrastructure (in terms of safe cycling routes and cycle storage) making the town – as one resident put it – 'the most unfriendly place to cycle I've ever cycled in'. Another involves the constraints imposed by the town's physical infrastructure. In many town centre streets, terraced houses front directly onto the street, with parked cars lining both sides of the road. Parking generates all kinds of conflict over the legitimate uses of public space: but the underlying constant is that 'too many demands are being placed on the finite carrying capacity of the environment' (Fisher, 2017: 8). Having observed these features of the town, we asked people in our second survey how they get around. The findings demonstrate that the car dominates mobility within Macclesfield (see Table 5.1). Our wider research has shown that this auto-dominance is today 'marked' as an everyday security concern; a matter that impacts detrimentally on people's sense of the town as a safe and liveable place.

This impact is, however, framed in competing ways. One framing treats problems associated with the car as simply a feature of the way the modern world is constituted. This is a sentiment we might call *auto-fatalism*. It is the overarching tone of many local conversations about the car-centric quality of the town. Sometimes this fatalism is explicitly stated. 'I think

Table 5.1: Transportation in Macclesfield by mode (2022 survey)

	Car (%)	Walking (%)	Bicycle (%)	Bus (%)
Uses daily	50	51	4	2
At least once a week	30	29	9	12
At least once a month	4	6	6	11
Seldom or never	15	14	81	75

where we live', one female participant observed, 'it's never going to be quiet in terms of traffic. And if I let myself be bothered by that, it would spoil where I live'. This remark is an apt illustration of Zerubavel's (2006: 9) point that denial takes active work, requiring 'a deliberate effort to refrain from noticing' and 'refusing to acknowledge the presence of things that actually beg for attention'.

More often auto-fatalism serves as a background assumption that structures what can be thought and said. On this view, cars are essential to how people live – and must live – today. This is taken to be a fact about the social world that it is beyond our capacity to alter. So too is the fact that people have more cars (it's their choice, and modern life anyway renders this necessary); that cars have become bigger (what choice do we have when manufacturers design cars that way?); that more journeys are made by car (what other safe or realistic options do people have?); and that cars are stored on crowded urban streets (they have to be parked *somewhere*).

On occasions, these claims are articulated in the idiom of personal freedom, even form part of an ideology of autophilia (Bentley, 2019): this has become a common motif in wider public contestation over urban transport (Schwanen et al, 2026). But this requires active consideration of the questions at stake that is for the most part lacking – and, as mentioned, in Macclesfield this outlook has not been provoked by any active policy agenda to restrain car use, or limit where they can be driven. More often this framing has an unthought quality, a 'psychosocial attachment to automobile autonomy' (Charbonnier, 2021: 251), and habituated sensibility towards the car's presence in the urban landscape, that 'sidesteps' questions of 'social accountability and blame' (Brewster and Mayerfeld Bell, 2022: 51). On this 'natural' framing, car-dependency reverts to being like the weather: one can chatter and grumble about it, but it is beyond the reach of social evaluation and imaginable interventions. Automobility is judged so 'intrinsic to society', and change would require such 'an extraordinary degree of public acquiescence and social mobilization' that the 'underlying institutional contexts are immutable' (Rajan, 1996: 74, 219). This kind of fatalism amounts, in the end, to a form of denial (Harrington and Shearing, 2017: 32). The car returns to the status of 'unmarked' (Zerubavel, 2018).

A second – more common and active – framing of everyday car concerns we might call *auto-othering*: the harm caused by the car is the responsibility of a minority of 'bad' drivers and 'careless' owners. On this view, as we hinted earlier, the mundane disorders of automobility are pinned on those who speed on residential roads or park their cars without care or consideration for others or needlessly make short journeys by car to drop their children outside school. This framing draws upon forms of responsibility allocation and moral boundary-drawing that are commonly associated with public discourse on crime. To borrow the comparison drawn by Garland (2001), road traffic risks are made familiar – and reassuring – by placing them in the same moral register as (other) crime risks. What follows from this framing is also familiar. Speeding, bad parking and cognate car harms are problems that call for more or better enforcement by police or other local authorities. Such action – as one complainant about speeding on Macclesfield's residential streets put it – is 'necessary to deter future idiots'.[9] On this 'us' and 'them' characterization, car troubles are rendered a risk created by 'others', and they can and should be tackled in ways that leave 'our' forms of living and moving unaffected.

Against this backdrop, auto-othering begins to look like another form of denial. Its police-centrism is a refusal of what (else) is required to seriously address automotive harms, especially car-restraining measures (such as street re-design, low-traffic neighborhoods, School Streets schemes, or congestion/emissions-charging) that may affect 'us'. As Appleyard (1980: 117) puts it: 'The city that relies solely on better enforcement as a solution to its residential traffic problems is avoiding the issue.' The auto-othering frame is also a refusal of authorship, a form of reassurance that local car troubles do not implicate 'our' way of life but are caused by 'them'. 'We have such a hard time', David Wallace-Wells (2019: 35) writes, 'acknowledging or understanding our own responsibility or complicity in the changes now unfolding, and such an easier time evaluating the morally simpler calculus of pure victimhood'. 'Complicity', he continues, 'does not make for good drama' (Wallace-Wells, 2019: 149).

This suggests a third frame: *auto-complicity*. This framing intimates that a wider set of questions flows from the observation that people in Macclesfield are today (in ways that they didn't a quarter of a century ago) *noticing* cars and *registering* their deleterious effects on the urban landscape. One dramatic exemplification of this larger framing took the form of a 'Death and Climate Change Sold Here' protest daubed on a filling station in a middle-class neighbourhood that was reported during our fieldwork.[10] That protest is one way of dramatizing – and drawing inferences from – a typically more muted realization that people are implicated in producing the automotive disorders about which they are concerned – they are co-authors of their own problems. 'Those who suffer from the harmful effects of automobility are', Rajan (1996: 218–19) writes, 'simultaneously bound up in the institution through which such risks are produced. There is no clear distinction between the producer of harm and its victims; most of us are both'. This acknowledgement is also focused on the felt necessities built into people's daily lives by car dependency: that we drive because it's become the only realistic way to get around; we drive because it's not safe to walk or cycle, but by driving we undermine the safety of other mobilities and bolster an urban infrastructure that privileges automobility. In other words, people feel trapped in their cars; coerced into freedoms that entail the ongoing production of a series of 'ordinary harms' (Agnew, 2020; see also Anker, 2023).

This wider framing is permeated by a 'greening' of worries about disorder. The forecourt protest is one activist instantiation of that altered ecology of security. But perhaps what is also going on, in the mundane concerns and grumbles about the place the car has come to occupy in the local harmscape of one English town, is a quieter reappraisal, and renewed contestation, of the car. That reappraisal suggests that the emotional palette of sensibilities towards cars is not restricted to gratification and attachment; cars and driving also ground equally visceral forms of judgement, indignation and censure (Sheller, 2004). It suggests that cars are raising fraught issues about the meaning and uses of public space. In local hot spots of contention, such as around school gates, cars loom large in people's censorious views of others' conduct, and their often-impassioned condemnation of drivers' recklessness, selfishness and lack of consideration. Beyond this, there are also detectable signs that the issue is being cast within an emergent ecological consciousness towards the impact of the car on the urban landscape, one that problematizes driving, speeding and parking as indicators of people's concern for, or carelessness towards, the quality and liveability of the (local) environment. If this is even partially correct, we may increasingly see long-standing concerns that have hitherto been construed overwhelmingly under the sign of orderly or disorderly conduct (Newburn and Ward, 2022) being reimagined as matters of planetary (ir)responsibility and entangled with the question of how to create liveable urban environments and sustain viable planetary futures.

On car-centric harmscapes

This account of some emergent features of local mundane culture in one English town cuts against – and problematizes – some preoccupations and omissions of existing scholarship where cars are concerned. It suggests that the relegation of questions of automobility, and its attendant risks, in criminology and security studies is, literally and figuratively, unsustainable (see, further, Loader, 2025). The notion that crime is typically, even inescapably, salient in public discourse, whereas cars and traffic are not, no longer looks universally plausible: motoring stands today on a faultline between the politics of reaction and possibility. The idea that social disorder is limited to noise, nuisance and damage caused by a minority of troublesome – typically young, poor, male – others, and cannot also be an effect of the use and enjoyment of a taken-for-granted consumer object by masses of ordinary people, starts to seem equally problematic. These seem like relics of older ecology of security.

We have seen, in this chapter, that cars are a matter of dispute. If our predominant orientation to the places and relationships that we inhabit is one of concern, as Sayer (2011) suggests, this helps us to understand why people find evidence of what they see as the absence of concern so distressing. Carelessness in driving in this context, for instance, means more than just a lapse of attention. Rather, it signifies something about egotistical self-assertion as a way of being (Mitchell, 2005), in a way that contrasts with an implicit demand, or at least hope, for something more sociable and more ethically engaged. These seem to be some of the ways in which people's expressions of concern about everyday social etiquette merge into their larger worries about the future of their habitat and its place in the social ecology of places.

We might usefully, in this regard, see towns and cities as in large part constituted by car-centric infrastructures and movements and investigate auto-dominance as an indirect cause, and direct agent, of urban disorder. If we do so, we might also confront and reconsider the puzzling criminological separation between public safety and road safety and the attendant tendency to conceptualize and act upon the latter differently, as a distinct and less urgent issue, rather than as a central question of the urban experience. When we think and speak of public space (and its meanings, uses and possibilities) we generally do not mean streets and roads. These have been lifted out of town- and cityscapes and treated as distinct socio-juridical sites (Braun and Randell, 2022) – places where motorized mobility has come to trump the demands of dwelling and suppress alternate possibilities for care, nourishment and sociality (Bertolini, 2020). Our research suggests it may be time to refuse those modern motonormative distinctions and, instead, rethink urban security in ways that acknowledge fully and address seriously the everyday harms of car-centric transport systems.

SIX

Ecological Governance of Place

In this chapter, we want to explore efforts that people make to care for and repair what they see as damage or threats to the town, its social fabric and physical environment. Some of these actions are quite small and undramatic in character and not all of them are likely to be seen as political acts in any overt sense. These forms of 'quiet activism' (Steele et al, 2021: 2) tend to involve 'modest, embodied acts of care, connection and creativity as part of everyday life, places and spaces'. Across Macclesfield, we observed many examples of practices of this kind – litter-picking, recycling, gardening, minimizing food waste and numerous other expressions of local social responsibility. At other times, or in response to a range of threats and challenges, matters take a more recognizably political turn, whether in the form of community organization and sometimes protest, or in the sense of demands lodged against institutions to perform better what people take to be their duties of care and protection. The relations between getting together to do things and demanding that authorities take actions is a complex and sometimes fraught one. But they each arguably represent what Valverde (2011) calls 'security projects' directed towards meeting a need or ameliorating a current or possible problem. While it is no part of our intention to erode the differences between these very diverse practices, there may be some value in thinking of them each as participating in aspects of the governance of the town and its ecological challenges.

In his book on *Ecological Security*, Matt McDonald (2021: 147) asks how institutions responsible for public safety and security 'might be encouraged to orient their actions towards ecosystem resilience and the rights and needs of the most vulnerable'. McDonald's concern is with international security. We take up his question here and think through what it might entail with respect to the local governance of security – or what we call the ecological governance of place. What follows is necessarily a sketch, an attempt to offer some empirically informed lines of flight into what local security governance might require and look like in a climate-changed world. As Blaustein et al (2024: 9–10) put it: 'Description and theory-building must precede problem-solving, evaluation, and even normative debates when it comes to understanding and governing new and emerging harm landscapes.'

The term ecological governance is not meant to suggest that police and other local agencies wholly or mainly re-direct their efforts to environmental problems – though it in part signals that some re-direction is needed, as we shall see. Ecological governance is not defined by the problems it seeks to address. Rather, the term signals an orientation to stewardship, to the care and repair of the social and physical and natural fabrics of place, to recognizing the entanglements between the local and planetary, and between human and more-than-human worlds. This re-orientation is likely to mean enhanced attentiveness to questions of climate mitigation and adaptation, to the issue of local resilience. But it entails, more capaciously, an altered approach to a wide range of other crimes, harms, disorders and vulnerabilities of the kinds we have detailed in this book. Ecological governance does not tell you what problems agencies have to think about; it is way of thinking about a whole gamut of problems entailed by the responsible governance of place. Let us explore how that changed orientation to governance is rooted in our research.

Environmental care-taking

People in Macclesfield do not merely respond to worries about crime, forms of social suffering, or harms in and to their environment, by demanding action from relevant authorities – though they do make such demands, as we shall see. People's investment in place also gives rise to various practices of active care and repair among residents, which we have observed, and whose value (and limits) it is important to understand. One local professional summed up the picture in the town as follows:

> There is quite a thriving third sector base in Macclesfield, there's lots of organizations who work together. There's a Macclesfield food partnership that's been set up, they offer different types of food provision, but they all work together so they don't duplicate, and they help each other out. The churches are very heavily involved in community work in Macclesfield, more so I would say than in any other area.

One sign of the changing landscape of concern and sentiment in the town concerns the fate of one such form of everyday citizenship: Neighbourhood Watch. In the mid-1990s when property crime was historically high and a key local concern, Neighbourhood Watch schemes were active in the town. We spoke to several Watch group members and coordinators in our first study. A quarter of a century on, we encountered very little evidence that Neighbourhood Watch remains a presence in the town – though some of the work it performed no doubt now takes place on local WhatsApp or Nextdoor groups. Other groups concerned with public safety are active in the town today. Street Angels is one example, though its purpose lies more in offering safe refuge and care for the distressed and vulnerable in public space on Friday and Saturday nights than controlling disorderly people. Age Concern is another, though its work in protecting elderly people from various forms of online crime (bank fraud, romance-scamming, and so on) is indicative of the ways the crime threat has changed over the last several decades.

These groups form part of what we found to be a rich network of voluntary action in local civil society concerned broadly with the everyday security of people and environments (see also Millie, 2022; di Ronco and South, 2026). This engagement took various forms. Concern about the condition and future of the town's public realm, and how to preserve or modify it in the context of diminishing local government resources and altered consumer habits, prompted institutional responses, as well as more diffuse forms of civic repair. The former included the post-COVID Macclesfield Town Centre Recovery Group, and several place-making, 'greening' and public art projects in and around the town centre, such as murals on key thoroughfares and around the railway station. The Town Council also set up a Town Ranger service in April 2019. The Rangers had an active social media presence and responded to public suggestions about locations where intervention was needed. The small group of Rangers cleared and cleaned road signs, difficult-to-access ginnels and other places that harboured forgotten rubbish. They also supported the town's numerous local litter-picking groups, carried out minor repairs to street furniture and assisted public gardening projects. More than one local councillor wryly observed that the Rangers were more popular than local politicians. The work of the Rangers was often cited as a notable act of institutional care, as in this active resident's account:

> Closed shops, obviously, are a concern, but at the same time, it's always nice to see new shops opening up or reopening; like Flour Water Salt for instance, is back open again. And I do think a lot of people are doing the best they can with what's going on at the moment. There are some landlords that are charging too much, but there are some at the opposite and are really egalitarian and helpful. Everyone blames the council for the rates, but they're set by central government, so it's not the council's fault at all for business rates. What else do I notice? I see the effort that Macc town council for instance puts in with things like Britain in Bloom and the work the Rangers

> do. Just on what might like look like a small scale trying to make the town more attractive.

Initiatives to protect the town centre and repair the public realm often blend into those suffused with wider planetary concerns. The town is replete with several responsible consumption initiatives, for example, including 'Buy local' campaigns and efforts to promote small independent businesses, as well as zero waste co-operatives, sustainable local delivery services, and repair and re-use initiatives (see, more generally, Schoolman, 2020). These speak to, and seek in small ways to build pathways towards, an alternate future for the town centre, focused not only on retail consumption but on the importance of civic rituals and green infrastructure. They are forms of active citizenship that seek a transition from 'me towns' to what Dobson (2017) describes as 'we towns'.

In the decades following our first study of Macclesfield in the mid-1990s a number of local groups emerged and evolved around environmental and sustainability issues. In 2009, Food4Macc was established with the aim of preserving knowledge about food and addressing high food prices, but also to promote the importance of consuming local food and supporting the town's businesses. It situated itself within the global movement of 'transition towns' established in 2006 to encourage grassroots community projects for self-sufficiency, and take local steps against climate destruction (Hopkins, 2008). Food4Macc helped establish food-growing community gardens in public spaces and local schools, as well as running a garden sharing scheme. It disbanded in 2022 due to prohibitive insurance costs.

By 2018, however, the activities and values of Food4Macc had been embraced by Macctastic, an umbrella, non-hierarchical eco-organization which now also encompassed work on climate, ecology, reduced litter and plastic initiatives, de-carbonizing transport campaigns, as well as a repair cafe. The origins of Macctastic reveal the dense interconnectedness of local action and sentiment. It was founded following a Treacle Market (a popular and very well-attended artisan market held in Macclesfield once a month) and Barnaby Festival (the main arts organization in the town from 2010 to 2024) collaboration in the form of a stall at the monthly market. The stall initially focused on the environmental impact of plastic and cutting plastic use, but its popularity and interest during the market led to the setting up of Macctastic. By the time of our fieldwork in the early 2020s, the town was rich in environmental groups, some acting under the umbrella of Macctastic, others emerging beyond it. These included Grow Macclesfield, the One Project and Macc Wild (which focused on allotments, community gardening, rewilding and recycling projects, annual green festivals, removing weeds and installing/renovating planters), as well as schemes for recycling one-use products, a community swap-shop, and food waste reduction initiatives. The presence of such groups was inscribed in the everyday material life of the town. Little signs marked sites where such groups had intervened to restore nature, grow food or remedy and make useable neglected corners of Macclesfield (both in the town centre, but also on outlying estates). The groups also had an active digital presence on the Town Council's social media channels and neighbourhood Facebook groups.

Third-sector organizations are also an established part of the ecosystem for the provision of essential services to vulnerable groups and, as such, form part of the 'mortar between the bricks of services' (Stewart and Eborall, 2025). But the institutionalization of third-sector groups within infrastructures of state provision belies the fact that their work mostly relies on people's sense of entanglement and relationality, and the ongoing commitment of individual volunteers. These groups make up a shadow infrastructure which is mostly evident in areas of high deprivation, but whose presence has today become vital in communities that do not typically understand themselves as marginal. Residents spoke to us about local organizations that provide myriad services and support for vulnerable groups (Food Pantries and Foodbanks, Cheshire Streetwise, Cre8 Youth and Community Programme, Just Drop-In, Space4Autism, Cheshire East Hospice, Grow Macclesfield – Community Allotment, Disability Information Bureau, Reach Out and Recover, as well as the aforementioned Street Angels). Some of these services have taken on responsibilities that in earlier decades may have been provided by the local state, most obviously in the case of youth work. Many we talked to were involved in volunteering or supporting the work of these organizations. We also observed agile responses to new vulnerabilities, such as Warm Space initiatives during the cost-of-living crisis in winter 2022–3 and new neighbourhood-focused provision emerging, such as the Weston Community Hub. The churches in Macclesfield were actively involved across many of these initiatives. Hope in NorthEast Cheshire served as an important hub of a dense network of community action and care. They provided a base, organizational capacity and volunteers for the Street Angels, as well as contributing to food waste and food bank initiatives (both during and after COVID-19) and supporting young people through Just-Drop-in and Cre8.

While the motives behind these 'communities of practice' (Wenger, 1998) were diffuse and differentially embodied in an 'ecological habitus' (Kirby, 2017), the thread that appears to connect their responses to local insecurities, vulnerabilities and fraying of local infrastructure is 'what needs our care and concern' (Honig, 2017: 28). Some of this activity in civil society involves making good on the absence or failure of local state provision under conditions of ambient austerity. But it is reductive to interpret it solely in these terms. We need, instead, to reconsider the relation of everyday security to practices of care for people and habitats (Jupp, 2022; Traill et al, 2024). These groups hint at a conception of ecological security focused on looking after the places we inhabit, rather than simply protecting them from external threat. Care as a disposition of concern and form of action is ambivalent and multivalent. Food pantries were promulgated as both a solution to a cost-of-living crisis and a local antidote to a global system of waste in food consumption: action for the community and the environment. Urban gardening addresses food poverty while providing a resource for vulnerable groups. Some of these projects engage elderly residents or recent refugees/migrants in a bid to tackle loneliness and isolation; they also actively support those with addictions. As one local organizer remarked, the community allotment provides care across social classes, from socio-economically vulnerable groups to those who may be feeling isolated across Macclesfield's affluent suburbs and satellite towns. Care for the social and material environment of the town goes hand-in-hand with wider environmental concerns and protecting the most vulnerable groups within local ecosystems.

Many respond, then, to felt concerns about the places they inhabit and the evident failures of public provision not merely by making demands of authorities, but by engaging in the 'dull, routine, everyday work' of care and repair (Gupta, 2021: 41; see also Williams, 2016). In response to environmental degradation, we encountered repertoires of dwelling, of cherishing, protecting, preserving and caring for (this) place – forms of civic engagement one might posit as green variants of 'collective efficacy' (Sampson et al, 1997; cf Lynch and Boggess, 2015). Much of the attendant action and reaction is concerned with boundaries (what is worth protecting and from whom, what is out of place), as well as the quandaries of stewardship, state and individual responsibility. The practice of security here does not mean seeking or living in a sanitized environment free of care, but an 'awareness of entanglement and relationality' (Harrington and Shearing, 2017: 132) that connects local acts to global concerns, and treats the unsightly, the out of place, and the abject as a relational failure and withdrawal of care. These mundane acts of urban care-taking, of everyday attentiveness and embodied 'making and doing' (Steele et al, 2021: 139), also suggest the need to think afresh about the connection between the maintenance of physical and social infrastructures engaged in by civil society and the work of statutory agencies tasked with supplying an infrastructure of protection.

Continuity and change in demands for governance

We conducted our original study of Macclesfield in the mid-1990s. Our fieldwork took place just before the return of a New Labour government in 1997, with its commitment to increase funding for the police and guarantee a dedicated officer for every neighbourhood. In the years prior to that reform, it had become senior police and governmental orthodoxy that generic visible patrolling was an antiquated police tactic, an ineffective use of scarce police resources. Senior officers at the time became fond of quoting Home Office research purporting to show that a police officer will expect to pass within 100 yards of a burglary in progress once every eight years (Clarke and Hough, 1984; Longstaff et al, 2015). Police resources, the argument went, were better spent in other, more targeted ways.

Against this backdrop, our 1990s research was replete with angry, visceral complaints from Macclesfield residents about the absence – or, as they saw it, disappearance – of the police from local social relations (Girling et al, 2000: ch 6). These concerns resonated through our interviews and focus groups and were loudly pressed at neighbourhood meetings we observed. They were expressed in several, overlapping, idioms. Sometimes concerns were couched nostalgically as a yearning for a 'lost England' imagined and recalled as a land of orderly, cohesive communities, at the material and symbolic heart of which stood the 'bobby-on-the-beat'. Others relayed a story of technologically mediated remoteness, wherein the motor car had mobilized, sped up and geographically diffused policing, and re-located officers at the end of phone, at great distance from the lives and concerns of the communities they served. For others, the chosen register was misalignment and detachment. The police, on this view, answered to internally generated, bureaucratic priorities at some remove from people's concerns about, and desire for, local guardianship. On more than one occasion during our research, we observed the local Chief Superintendent explain to public meetings why visible beat policing was not his priority, only for the assembled residents to leave angrier than they had been upon arrival. It was as if an old, once trusted, friend had let them down:

> All you get told when you ask about the beat bobby: 'Well do you want less burglaries, less car crimes, or do you want them to always be on your estate and have problems with you'. It's sort of prioritize and all this, we just get nowhere. You saw what happened at that meeting. … I know they need more money, they need more police, but surely somewhere they can do something about it. I don't think they realize how worried people are by what's going on. I don't think they realize how it affects people's lives. (Quoted in Girling et al, 2000: 122)

Scroll forwards 25 years. The contemporary forms of everyday harm described in Chapters Three to Five also prompt concern because they signal an absence of effective or responsive governance in the town. The police remain a significant dimension of this feeling that the town is not being properly governed, and important to people's sense of secure belonging; indeed, analysis of our first survey found a strong positive association between trust in police fairness and people's sense of place (Bradford et al, 2024). Asked in our survey how often they 'saw a police officer or PCSO on foot in your local area', 39 per cent reported 'at least once a month' in 2021, a figure that fell to 26 per cent in 2022. Yet though the issue of police visibility came up from time to time in our interviews and discussions, as well as on local social media, the issue attracted nothing like the levels of visceral anger and disappointed exasperation that we encountered in the town during the mid-1990s. This could be the product of grim fatalism, an effect – and affect – of durable austerity. This structuring context may have produced an entrenched feeling that 'they' never listen to 'people like us', or that policing is never going to get any better. A frequent refrain in accounts of the police in focus groups was the resigned feeling that 'they just don't have the resources', rather than a call for accountability. Some worried in this respect about 'a sense of defeatism' and remarked on the dangers of the police 'giving up on things' (male resident).

But it also often seemed that 'calling for more police' had morphed into a thing that people say (after all, who could be against that!), without anything approaching the same level of intense feeling that had once animated that demand. People spoke of the reassurance supplied by seeing 'key workers in the community, even when there isn't an emergency', or wanted to 'see the police talking to children in schools', as two female residents put it. Some parents of young children reported that the only place they saw the police was when officers were regulating traffic and parking at the school gates. But there was something very low-key about such demands – it seems not quite right to even call them demands. Sometimes there is recognition that much of what the police now do to protect the vulnerable from exploitation and serious harm takes place out of the public gaze. As one retired resident reflected:

> It's like architects are paid to understand what's going on in the bits of the building that are unseen. Perhaps the most important part of the job. I think it's the same with the police. A lot of their most important work, the general public aren't to know, and they shouldn't know. That's policing.

Residents often expressed positive support for the police. When asked to rate the performance of the local police in our 2022 survey, 36 per cent said they were doing a 'good or excellent job' while only 11 per cent believed them to be doing a 'poor or very poor' job (the rest gave a rating of 'fair'). When asked to identify the three things they wanted the local police to prioritize, 'drugs and drug-related crime' ranked highest (34 per cent of first choices), followed by 'anti-social behaviour' (18 per cent) and 'traffic/road-related issues' (14 per cent). Tackling domestic abuse was fifth (6 per cent of first choices). None of this is to say, however, that we never encountered participants making a case for greater visible policing. The following is a case in point:

> I'm not advocating to have a police officer on every corner, but I do think the police fulfil a vital function. I think there is a lot of crime, and I think if the police were more visible it would create more a sense of environment. As I say, as a male, it's easier for me to feel safer, but if you're an ethnic minority, if you're a woman or if you're a younger woman, you aren't going to feel that same sense of safety. And that's where I think visible policing is so important in a general sense but also creating the sense that there are police going around, so I shouldn't park here, or I shouldn't shout racist abuse at somebody. (Male resident)

To focus on such framings would, however, represent a failure to attend closely to important aspects of our participants' views. Many of them would support firm police action being taken to discourage drugs and other forms of anti-social behaviour. They want the town, or their part of it, to be defended against incursion (see, more generally on this, Bradford et al, 2025b). But the question of sanctions for deterrent or corrective purposes did not predominate in our conversations; sometimes whole group discussions would sail by without anyone raising matters of police. These discussions tended to focus rather more, as we have seen in Chapters Three and Four, on the condition of the local townscape, and of sensitive places within it, such as the town centre or public parks; and on the whole they are more focused on pleas for investment than on demands for visible policing. Residents also tend to acknowledge that many local security problems – in respects of such matters as litter, or traffic pollution, or the decline of the town centre – are in part of their own making.

This in turn prompts questions to do with care and custodianship rather than simply of enforcement. Sometimes, people were quite explicit in attributing what they see as the degradation of the local environment to an ongoing politics of fiscal retrenchment. 'Their budget is being slashed since the age of austerity', one male participant observed, 'Councils are getting hammered for central government policies and that's why services aren't as good as they should be.' More generally, participants raise questions concerning responsibility and worry about an abdication of care and guardianship on the part of local authorities, a failure of government properly so-called, not simply of policing.

Much of the concern about the felt neglect of place is, in other words, directed today at local government, which is commonly perceived to be absent, or failing in its core tasks; to the extent that local government has come to occupy the symbolic place (as the vital but absent guardian) that was held by the police in the 1990s. Some of these concerns were peculiarly local. We found confusion about who held relevant responsibilities and budgets for services in the town (Macclesfield Town Council or Cheshire East Council), and widespread disquiet about Cheshire East Council especially. In our second survey, 34 per cent of participants thought Cheshire East was doing a 'poor or very poor' job, compared to 25 per cent for the Town Council. This may reflect the fact that Macclesfield Town Council, despite its limited powers, was felt to have sought to address the quality of neighbourhood life through small acts of place-making and funding events to encourage residents to use the town centre and take pride in their locality.

In our interviews, Cheshire East Council was, by contrast, variously described as 'inactive', 'remote', 'not caring', 'dishonest/corrupt' and 'not responding to people's priorities'. One woman complained about a lack of engagement with local people and civic groups: 'It's very paternalistic, very top-down.' Many spoke – in the spirit of this male resident – of 'not having much faith in the Council to do the right thing … I'm not sure they care enough about people that need help'. Common sources of concern were recent planning decisions, poor maintenance of road surfaces, the lack of public transport, failure to maintain key physical and social infrastructure within the town, and an inadequate response to mitigating the climate crisis.

Concern – and sometimes anger – was in these ways directed at specific local authorities and their felt failure to exercise a duty of care towards place. People were angry about local authorities that have been starved of resources and appear remote and disconnected, in much the way the police were in the early to mid-1990s. But there is also something more fundamental at play here. What people were noticing in their everyday lives, and bemoaning in conversations with us, was the absence of local governance as a structuring feature of social life. It is the absence of an entity that is concerned with, and cares for, the basic infrastructure of place that underpins not only the safety but more generally the quality of neighbourhood life, and the wider environment. It is, as we argued in Chapter Four, the absence of attention towards what the political theorist Bonnie Honig (2017) calls 'public things' – neighbourhoods, a town centre, parks, play-spaces, roads, libraries, flood protection, and so on – things that 'we' share and steward (and contest) in common as concerned inhabitants of a place, and as democratic agents. As one of our participants put it:

> Well, I'm coming at it from a different perspective, in that, as I say, I don't feel unsafe in Macclesfield. And I don't know, therefore seeing police on the beat wouldn't make

> that much difference to me personally. I think if I was making decisions on how to spend public money, I'd put money into preventative services or public services or something like that. Street cleaning, urban space stuff, but really, I think, a lot of the problems in society in local communities, come from a lack of investment in public services. Or cuts in public services year after year. (Male resident)

One way of re-casting these concerns is to understand them as demands for forms of security governance attuned less to law enforcement, sanctions and othering, and rather more to the care and repair of place, and to the exercise of responsible environmental guardianship, broadly understood. Such demands are in keeping with the practices of urban care-taking engaged in by many of the town's residents. How though can these informal practices and inchoate sentiments be fashioned into an altered vision of ecological governance? What 'immanent processes and tendencies' (Amin and Lancione, 2022: 3) do they reveal for reconfiguring how we frame, and seek to realize, everyday security today? Let us try to sketch out some key elements of such generative possibilities.

On the ecological governance of security

A small body of criminological work has emerged in recent years concerned with how climate breakdown is reshaping – and is going to affect further – policing and associated agencies of urban governance. This work takes various forms. Some of it maps future climate risks and the challenges these are likely to pose for the police (Matczak and Bergh, 2023; Matczak, 2025) or, conversely, detects signs of emergent new practices of private governance and elite secession, so-called climate-gating, or 'bunkerization' (Lam et al, 2022; Berg and Shearing, 2024). Some research documents empirically how police are adapting to actual climate emergencies, such as bushfires (Blaustein et al, 2023) and flooding (Mutongwizo et al, 2022), or explores the likely roles and potential of new constellations of polycentric urban management (Malik and Berg, 2024). A new paradigm of 'resilience policing' is slowly beginning to take shape (Mutongwizo et al, 2021). For present purposes, this small body of work offers some resources for analytical re-description, a jumping-off point for the task of bringing an 'ecological imaginary' (Gandy, 2022: 31) to the governance of place.

Let us begin with that emergent paradigm of 'resilience' or 'adaptive' policing. Proponents of this model start from recognition of new, uncertain, 'situated and yet always translocal' harms (Amin and Lancione, 2022: 3), whether in respect of cybercrime, terrorism, illicit global markets in trafficked people and goods, or climate change-induced disasters, such as heat, fire or flooding (Mutongwizo et al, 2021: 611). Their preferred response to this altered harmscape is to bring the 'mentalities, functions and capabilities of policing institutions and networks' (Blaustein et al, 2024: 4) into alignment with broader forms of security governance in the Anthropocene, in order that they can operate effectively across the temporalities of what Crawford (2018) calls 'everyday', 'emerging' and 'existential' risk. Diverse policing capacities are needed to anticipate and respond to threat in this environment, and, on this view, a core police task is to 'enrol other actors, for example, government and community resources, to deal with these harms' (Mutongwizo et al, 2019: 611). The police need to become 'facilitators/enablers in community-capacity building' with a view to creating forms of nodal governance that are actively distributed and shared across the community and police. 'The role of the police', Mutongwizo et al (2019: 614) contend, 'does not only involve practically dealing with these harms but also includes bringing together the resources required to restore order, bounce back following an incident, as well as honing response capacities for future events'.

There is much to commend in this proposal to rethink what policing is to be in a climate-changed world. It alerts us to novel threats to socio-ecological systems under conditions of climate breakdown and seeks to 'encourage practices oriented to sustaining their conditions of existence' (McDonald, 2021: 112). It is focused on the question of how to manage the disruptive effects of catastrophic risk in 'an equitable, efficient and responsive manner that expresses an ethics of care' (Blaustein et al, 2024: 7). We build upon these elements in what follows. But there are also reasons to think that resilience policing is insufficient as a basis for thinking through what is meant and entailed by the ecological governance of place. Some of these reasons have to do with the idea of 'resilience'; others relate to the paradigm's police-centrism.

There is a vast literature on resilience that tracks its journey across the natural then social sciences, to governmental rationality and therapeutic jargon, that we lack the space to engage with here (Olsson et al, 2015). The core meaning of the idea remains nonetheless clear: it refers to the ability of an individual, community, organization, society or system to recover, or 'bounce back', from a harmful shock. In socio-ecological contexts, resilience means the capacity of a system to 'absorb a disturbance' and 'still retain the same function, structure, identity, and feedbacks' (Walker et al, 2004; cited in Mutongwizo et al, 2019: 606), or, to use the definition crafted by the Intergovernmental Panel on Climate Change, 'the ability of a system and its component parts to anticipate, absorb, accommodate or recover from the effects of a hazardous event in a timely and efficient manner' (2012: 5). Certain key aspects of this idea limit its value for thinking about everyday socio-natural harmscapes.

One is that resilience policing remains emergency/disaster focused; it is oriented to preventing, preparing for and responding to serious, even catastrophic, risk. This is obviously one component of what the ecological governance of place has to entail today. But that mentality and its attendant capacities seem less suited to thinking about, and acting upon, the kinds of slow-burning public safety concerns and environmental harms that we have described in this book. These everyday disorders do not inflict one-off damage. They entail harm that accrues over time, that suffuses everyday life on a recurring basis, and that often lacks identifiable culprits. These are, rather, chronic harms that flow from entrenched habits and valued ways of life, and which are differentially co-produced by agents who are simultaneously, if we retain a criminological lexicon, 'offender' and 'victim'. The tenets of resilience policing are not the best guide to how we should frame and respond to these forms of everyday environmental disorder.

This catastrophe mindset is also central to a further limitation of resilience policing. The organizing mentality at play here is one concerned to fend-off disaster, adapt to looming threat and respond and recover if and when the worst – or something very terrible – happens. This is undoubtedly important. But the result is a governing disposition that tends to be both gloomy and conservative. Its orientation is to head-off emerging and potentially existential threats and to conserve in place extant communities and ways of life – even if those ways of life have proved to be harmful, even self-destructive. An ecological governance of place has to be about more than this: it needs to supplement a protective focus on resilience with a concern to think afresh about alternative and better ways of dwelling in, and moving within, habitats. It has to connect policies geared to mitigation and adaptation to a more hopeful politics of just transitions and transformation (Schwanen, 2021; Horne et al, 2024). It has to be about socio-ecological betterment, not simply survival.

The emergency orientations of resilience policing are also, arguably, related to its final limitation as a lens for thinking about the ecological governance of place: that of police-centrism. Despite their clear commitment to nodal practices of security governance, Mutongwizo et al locate the *police* at the organizational (and, by extension, ideological?) heart of resilience *policing*. 'In policing to enable resilience', they insist, '*police* act as facilitators and enablers who identify, and build, the capacities required for communities to effectively anticipate and respond to material shocks' (2019: 607; emphasis added). This focus makes sense at moments of emergency response when the unique coercive capacities of the police are both essential and legitimate (Bittner, 1990). But the more resilience policing is couched as a 'whole-of-society' enterprise (Berg and Shearing, 2011), the more difficult it becomes to justify police-led organizational forms. Making the police focal to resilience policing *tout court* risks securitizing the governance of security and extending a 'bandit-catching' mindset to governance tasks that stretch well beyond policing. It also risks dragging the police into ambient preventive and problem-solving work that falls well beyond their skill set and mandate (Loader, 2006). If ecological security is a 'whole-of-society' undertaking (and there are good reasons for wanting to make it so), the appropriate institutional frame for that undertaking is democratically constituted (local) political authority, not the police. This, perhaps, is the ecological insight about responses to everyday disorder that can be discerned from our participants' sentiments about the importance of holistic and attentive urban governance, and their anger about its careless absence or shortcomings.

So, the ecological governance of place needs to harness the insights of resilience or adaptive policing, while extending beyond it. But how? What (else) should ecological governance seek to do or be? By way of conclusion, two possible lines of flight are worth sketching.

First of all, it requires institutions and practices of governance that understand their object of attention and concern, not simply as people, communities and places as conventionally understood, but as socio-ecological orders. This in part means recognizing that such orders are 'entanglements of bodies, nature, things, technologies, infrastructures, and institutions'

(Amin and Lancione, 2022: 1). It also requires new apprehension of the altered harmscapes that need attention, and intervention (Berg and Shearing, 2018). Those harmscapes will continue to include person-to-person crimes, harms and vulnerabilities – the 'taking and hitting' offences that are the conventional staple of police and public safety authorities, offences that are increasingly mediated by translocal digital technologies. But the scope of attention needs to expand to encompass harms inflicted by persons on non-human entities: the built environment, the physical landscape, the sensory quality of urban atmospheres. It further needs to bring within the ambit of urban governmental attention a host of thing-to-human harms, whether in respect of climate change-induced risks of flooding, heat, drought or fire, or the routine dangers and environmental harms caused by car-centric mobility systems.

Attending to these harms as elements within a socio-ecological order may also require shifts in established ways of thinking and its associated organizational forms. It may, in particular, mean softening some of the mental distinctions, and abandoning certain of the categories, that structure the governance of everyday security. In the altered harmscapes we have described, one needs to ease up on any clear demarcation between everyday harms that are criminal and those that are legal but a source of annoyance and trouble, especially if that distinction is used to artificially select certain forms of 'anti-social behaviour' for intervention, while relegating or dismissing concern about others. It may be equally unhelpful to uphold a sharp distinction between those harms and crimes that get marked and responded to as 'environmental' and those which are assumed to entail some other kind of damage, or to draw too sharp a mental and organizational divide between questions to do with the quality and safety of neighbourhood life and those of climate change risk mitigation. Acting to produce or sustain the goal of public protection in ways that treat safety on the roads as some discrete part of the urban experience also seems increasingly untenable.

Some more concrete indications of what all this might entail can be gleaned from our research. It might mean, as we suggested in Chapter Three, responding to some 'old' forms of social disorder – vandalism, litter, fly-tipping, drug-taking, abandoned buildings – not simply or mainly as having individual victims, or as harbingers of serious crime, but as forms of ecological degradation – harms that first and foremost demand care and repair of the affected environment. It may, conversely, mean remaining attentive to how problems that manifest in the immediate environment may also be signs of exploitation, neglect or abuse in the wider ecosystem. Often, we found that young people hanging around in groups were treated by authorities and adult residents like this – less 'kids being kids' and more as a surface indicator of hidden vulnerability. It may also mean attending closely to the idea that 'normal', routine behaviours can be a source of environmental harm and experienced as disorder – as in the case of the various car-related troubles described in Chapter Five; as well as taking seriously the way in which state actions and inactions can be a source of felt incivilities – or, better, incivicities, damaging to the civic realm and the possibilities of civic association. Failing to protect valued physical habitats (such as peatbogs) is one case in point, as is failing to attend to the infrastructural effects of new housing developments. The inability to exercise proper care of basic infrastructure, or make provision for basic common amenities (parks, transportation, and so on) might also be interpreted as shortcomings of ecological protection.

The ecological governance of place is not solely concerned, however, with these reappraisals of the referent objects of everyday security. It requires, secondly, a revised ethos, or what Valverde (2014) calls a 'logic' – a particular set of aims and assumptions and their attendant aesthetics and affects. That logic is best captured by the regulative ideal of ecological urban stewardship (cf Eckersley, 2004). By this we mean practices of governance that are oriented to the harmscapes we have just described with an overarching disposition of care and repair, an aspiration to create, sustain and protect liveable environments, socio-ecological orders in which human and non-human life is able to flourish. It is an ethos that recognizes and works with an idea of security founded on entanglement and relationality between humans and non-humans and centred on the care-full protection of those who are most vulnerable within any given socio-ecological order (Harrington and Shearing, 2017). Intimations of this ethos can be discerned in the quiet practical acts of concern for vulnerable people and places we observed in our research, as well as in the distress residents of one English town evidently feel towards the absence of care-full authorities in their everyday lives.

In cashing out the idea of care-full and careful stewardship, some further contrasts can be drawn with resilience policing. Resilience often seems like a managerial ideal. It suggests a form of top-down, professional intervention conducted by those who are trained to appraise risk and organize effectively to prevent or respond to it. It all feels very monological. Dryzek and Pickering (2019: 35–6) explicitly contrast those aspects of resilience with an idea that feels much closer to stewardship: what they call 'ecological reflexivity'. Such reflexivity involves: recognition of the conditions and changes in socio-ecological systems, something that encompasses attentive listening to, and monitoring of, the habitats one is seeking to govern; active reflection on past experience, core values and possible interventions; and then, responding accordingly. We might view these components as part of 'an ongoing and iterative process' of environmental stewardship (Dryzek and Pickering, 2019: 36). Crucially, they have a close, arguably internal, relation to questions of inclusive deliberation and contestation (Dryzek and Pickering, 2019: ch 6; see also Eckersley, 2004: ch 6). Involving all affected in practical experiments in democratic governance is one vital means of enacting care and carefulness. The challenge is one of experimenting with how forms of 'caring democracy' (Tronto, 2013: chs 6–7) can best be given practical shape.

An ethos of urban stewardship also helps us to reframe the role and limits of policing as managers of socio-ecological order. We are very familiar with the idea of police as agents of order maintenance. At its worst, and all too commonly, order maintenance policing has been a polite term for street-sweeping, for the removal of marginal populations from the sight and minds of urban majorities. But as Matt Lloyd-Rose (2023) acutely observed in his immersive observations of police work in south London, much urban policing also involves routine micro acts of care, often grudgingly performed by officers who typically disavow this dimension of their daily tasks. To reframe policing as part of the infrastructure of urban stewardship is to make a virtue of that disavowed practice. It is to emphasize policing as a form of street-based repair (cf Smith and Hall, 2017); police attend to breakdowns in the surface of social order and extend protection, kindness and care to those who are structurally or situationally most vulnerable, whether in public space or behind closed doors. This is a practice that is necessarily coercive on occasions, but which at its best is part of the ongoing maintenance and repair work that makes urban life possible (Graham and Thrift, 2007; Hall and Smith, 2015). Policing, understood in this way, becomes a vital but limited aspect of practices of governance that see the overarching task as maintaining environmental security, by which is meant attending to the safety and well-being of all humans and non-humans who inhabit a place (especially those most vulnerable within it) and protecting and sustaining the physical and natural settings that enable those inhabitants to live a passably secure existence.

We are conscious that this is a sketch, if that. It is an outline of an orientation – or re-orientation – towards the governance of security, not a policy programme or set of guidelines or institutional arrangements (cf McDonald, 2021: 97). As such, it raises more questions than answers – to do with such matters as scale, jurisdiction, authority and organizational form (Valverde, 2014), not to mention political purchase and feasibility. We know this. But such issues we leave to others or at least defer to another day. What we hope to have prompted here is something prefatory to such detailed work. It is an exercise in imagining how one might conceive of public safety in ways that move beyond the important but limited question of how to prevent or police environmental crimes and ask, instead, what it might look like to act upon everyday security concerns in distinctively ecological ways.

Conclusion: On Earthly Security

We began this short book by revisiting some ideas of 'ecology' and 'environment' and considering how they had been used first in their source disciplines of natural history, and its successor environmental biology. We did this in order to look again at the uses of these terms in urban studies and in criminology and to offer a degree of reappraisal of the particular forms that they have assumed in those traditions. We noted that, for all our admiration of its many insights and practical contributions, the legacy of environmental criminology has some distinct limitations when it comes to thinking ecologically about humans' relations to their habitats and their capacity to thrive sustainably within them. We argued that despite their lexical similarity environmental criminology as a tradition, and more recent green criminologies, have developed in very separate ways and that this has distinct conceptual and, as we now want to argue further, political and ethical consequences.

Seen from within the horizons of criminological and social theory, it struck us as constructive to begin to develop a case for convergence between these bodies of thought. That would involve, as we put it, a certain greening of environmental criminology, in such a way as to reconsider aspects of human embeddedness in socio-natural environments, not only as sources of criminal opportunity or prompts to fear, but also as implicated in their ecological viability. The other face of this convergence, we suggested, entailed a development in 'green' criminology towards what are traditionally regarded as somewhat 'micro' concerns with everyday circumstances of living, in particular places, explored with a certain density of ethnographic attention. We argued that such a convergent approach might be helpful in thinking again about questions of security in mundane situations in a world where even the experiences of relatively privileged people, living in a relatively obscure and protected place, in a rich Northern country, take place within a penumbra provided by environmental degradation and climate breakdown. To the extent that it is necessary to name such changes of perspective, we described this as an effort at 'ecologizing security'.

Whether any of this causes anyone to rewrite a few paragraphs in a criminology textbook, or to nominate a new sub-field within security studies, is not really the point, nor our concern. We found it necessary to reflect upon these questions because they helped us to interpret certain things that people living in our place of study were telling us today that struck us as new and different from the problems that troubled people in the same town, and preoccupied British social and political life more widely, in the 1990s. Many of those things drew connections between the ability to live securely in a place and concerns over the condition and future of its environment; and while many of those concerns had very particular local aspects, none of them came without the shadow of planetary questions.

'If we are serious about even political life', Raymond Williams (2006: 116) once observed, 'we have to enter that world in which people live as they can as themselves … within a whole complex of work and love and illness and natural beauty'. We have set out to describe 'that world' as experienced by people living in one English town, and to understand their daily troubles and concerns against the backdrop of two aspects of the changed climate in which they live: enduring austerity and profound ecological mutation. In so doing, we have encountered concerns about disorder and security whose substance and mood seem importantly different from those we discovered in that earlier study of the same town, conducted in the mid-1990s. Concerns were expressed about a range of everyday disorders, some of which involved illegality, others of which were harmful but lawful. These disorders were viewed not as harbingers of serious crime, but as indicators of environmental degradation. People worried that things they care about, share and contest in common – parks, libraries, community centres, the town centre, and so on – are being left to decay. They wanted their streets and neighbourhoods to be safer. But often this means making public spaces better. They worried that the everyday security concerns that troubled them were – in part – of their own making. 'The problems follow you in your footsteps', Nikolaj Schultz observes, 'due to your own mode of life, your own mode of being, your own daily life' (Elmstrøm Lauritsen and Schultz, 2026: 6). Many spent time quietly looking after vulnerable people or neglected places, countering social abandonment and decay with small acts of repair.

Paulo Gerbaudo (2021: 9) contends that 'contemporary politics has become a politics of protection', organized around notions of sovereignty, control and security. As mentioned at the outset, such politics is often today cast in ways that drive a wedge between protection of places (homes, neighbourhoods, nations) and protecting the planet. The former is often imagined to be parochial, full of worries about the cost of living, boarded-up shops, GP waiting lists and lack of affordable housing; anxieties commonly funnelled nowadays into hostility towards foreign migrants and asylum-seekers. The latter, by contrast, is portrayed as a minority pursuit, concerns grounded in the lives of the comfortable, distant from the nagging worries of daily life at the bottom of socio-economic hierarchies. Centrist governments today struggle to straddle both of these horses, while a seemingly ascendant right-wing populism – fuelled by rage fostered in windowless digital spaces – seeks to cement the divide (Beveridge et al, 2024). An updated edition of the authoritarian playbook stokes and racializes fears of local invasion and decline, while actively dismissing the prospect of climate breakdown and castigating those who are concerned to stem it.

We've spent plenty of time over recent years analysing social media in Macclesfield and have sometimes encountered elements of this discourse. But we've also devoted countless hours to talking to people at length about their lives – on Zoom, in their homes, in civic spaces, or while walking or cycling with them, while also engaging with local groups and observing the quotidian routines and rhythms of the town. We have of course done this in just one place, and not everywhere is Macclesfield. But the town bears a family resemblance to many other places across many dimensions, in England and maybe beyond. 'Writing about a place', Rebecca Solnit (2014: 1) remarks, 'is to acknowledge that phenomena often treated separately – ecology, democracy, culture, storytelling, urban design, individual life histories and collective endeavours – co-exist'. So it has proved in the present case. By 'starting where people are' (Lucas and Reid, 2025) and paying close attention to what they notice, and the things they worry about and attend to as best they can, an altogether different picture emerges. It is one that appears to refuse the right-wing demand to choose between caring about home, place and community and expressing concern for the planet, and give insular, xenophobic priority to the former.

But to what kinds of 'projects of making more livable worlds' (Whatmore, 2002: 177) does our work speak? It speaks to the value of projects that acknowledge connections between the effects and affects of living under persistent regimes of austerity and those flowing from the climate emergency. It speaks to the value of interventions that attend to enhancements of human and non-human life in place, in the here and now, while cultivating action to sustain futures that extend beyond here, and beyond now. It looks to projects that link concern about everyday vulnerability – towards people experiencing neglect and exploitation and to landscapes that are degraded or threatened – in ways that stretch far beyond the 'law-and-order' frames that prevailed in the 1990s, and the narrow preoccupations of those who seek to re-heat or turbo-charge that politics today. It speaks, in sum, to an approach to everyday disorder that folds such problems within programmes to invest in the 'basic provisioning' and social and physical infrastructures of the 'good city' (Amin and Thrift, 2016: 129).

This too is a politics of protection, but of a different kind – one that acts upon the basic desire for safe 'dwelling places' by engendering connections between ecological and social questions (Latour, 2018: 53). It is a politics of security, but an earthly security calling upon new, altered practices of social imagination (Huijbens, 2021; Matthews, 2023). Such politics requires the creation of assemblages that do not yet exist and the formation of a 'we' that can affectively and effectively get behind it (Mouffe, 2022). It requires ways of governing that extend beyond demanding that the same old institutions pull the same old levers and which seek, instead, to craft societal mechanisms and principles of solidarity oriented to ensuring the sympoietic flourishing of the 'human and nonhuman agencies that constitute the particularities of a given place' (Matthews, 2023: 113–14; see also Haraway, 2016: ch 3; Marres, 2026: 21–3). This revised vision of what it might mean to live comfortably and collectively with risk may currently seem remote or far-fetched. We may only have been able to offer the briefest sketch of what it might entail. Yet it somehow seems close to the temper of the everyday security concerns we encountered revisiting this one English town.

Notes

Introduction

1 https://www.ipsos.com/ipsos-mori/en-uk/ipsos-mori-issues-index-november-2021

2 These were: climate change, coronavirus, Brexit, health, economy, poverty/inequality, lack of faith in politicians, immigration, education, housing. See further: https://www.ipsos.com/sites/default/files/ct/news/documents/2021-11/Issues%20Index_Nov21%20CATI_v1_PUBLIC.pdf

3 https://www.ipsos.com/en-uk/immigration-remains-biggest-issue-facing-britain-0. There is, however, contrary evidence that high levels of quiet public support exist for climate mitigation measures that is routinely under-priced by politicians. See, for example, https://ourworldindata.org/climate-change-support and Massey-Chase et al (2026).

4 One small indicator of this point can be found in a council by-election for the Macclesfield Central ward held in November 2025. The Green Party candidate took the seat from Labour with 41.8 per cent of the vote, compared with 28.6 per cent for Labour and 15.8 per cent for Reform UK.

5 Analysis of aspects of the findings can be found in papers on the following topics: mobilities across unsettling places (Bahceci et al, 2023); policing and sense of place (Bradford et al 2024); local urban disorder (Bradford et al, 2025a; Girling et al, 2026); and automotive disorders (Loader et al, 2025). An initial theoretical prospectus for the study can be found in Loader et al (2023).

two New Ecologies of Security

1 The UK government's assessment of the threat to national security posed by global biodiversity loss and ecosystem collapse can be found at: https://assets.publishing.service.gov.uk/media/696e0eae719d837d69afc7de/National_security_assessment_-_global_biodiversity_loss__ecosystem_collapse_and_national_security.pdf

three Socio-Natural Harmscapes

1 https://www.facebook.com/groups/193858551372087/

2 https://savedanesmoss.com/

3 https://www.cheshireeast.gov.uk/pdf/environment/env-health/broken-cross-aqma-map.pdf

4 https://www.cheshireeast.gov.uk/planning/planning.aspx Re Application 19/309M.

5 https://macclesfield.nub.news/news/local-news/save-macclesfield-greenbelt-launches-open-letter-on-new-chelford-road-homes. Ella Kissi-Debrah died in London in 2013 aged nine as the result of an asthma attack. In 2013 Southwark Coroners' Court found that air pollution had 'made a material contribution' to Ella's death. This was the first time in the UK that air pollution was officially recorded as a cause of death. https://www.bbc.co.uk/news/uk-england-london-68068828

6 https://www.cheshireeast.gov.uk/planning/planning.aspx Re Application 19/309M.

four Erosion of Public Things

1 Macclesfield has, it should be noted, seen a small growth in independent craft shops over recent years. The Market Place in front of the town hall has also, since 2010, hosted a monthly 'Treacle Market'. Both tend to attract Macclesfield's more prosperous consumers, as well as visitors from out of town.

2 These signs of decline were perhaps intensified in 2020 by the winding up of the town's football club – with its 147-year history of local pride and affiliation – following several years of gathering financial problems. Macclesfield proved relatively fortunate in that the club was quite soon recreated following new investment by a local businessman with a strong commitment to the town, and the enthusiastic involvement of ex-Premier League footballer, turned BBC pundit, Robbie Savage (https://www.placenorthwest.co.uk/macc-plots-stadium-improvements-after-takeover/). In January 2026, the revived club knocked Premier League team and Cup holders Crystal Palace out of the FA Cup.

3 https://macclesfield.nub.news/news/local-news/ryle-street-sinkhole-song-rocks-macclesfield-after-700-days-of-inaction

4 https://www.youtube.com/watch?v=t9x_B24hJUg

5 https://www.margaretthatcher.org/document%2F106689

five Driving Disorder

1 The same has arguably happened to the discursive place of the weather, and for similar reasons relating to heightened consciousness of global warming.

2 No local figures are available on the changing number of cars in the town. But the national trends are striking. In 1965, there were

7.7 million registered cars in the UK. In 1995 (the period of fieldwork for our original study), there were 21.4 million. In September 2023, that number of registered cars in the UK had reached 33.6 million. Save for the building of the Silk Road in the early 1990s, Macclesfield's road infrastructure has remained largely unchanged (Bahceci et al, 2023). Forty-five per cent of adult Macclesfield residents have access to one car/van; 27 per cent can access two or more cars/vans; 8 per cent have access to three or more cars/vans; 21 per cent have no access to a car/van. The equivalent figures for England and Wales are 45 per cent one car/van, 33 per cent two or more cars/vans, and 22 per cent no car/van (all figures from 2021 UK Census).

3 https://schoolstreets.org.uk

4 PCSO stands for Police Community Support Officer. PCSOs are a category of staff with a more limited range of powers than police officers and whose role is chiefly focused on local patrol and community liaison.

5 https://www.macclesfield-live.co.uk/police-shocker-officers-catch-speeding-15866814

6 https://www.racfoundation.org/media-centre/cars-parked-23-hours-a-day

7 The average weight of new cars in the UK has grown by 25 per cent in just seven years, from 1,553kg in 2016 to 1,947kg in 2023 (Martin, 2024).

8 https://m.facebook.com/Dangerous-Macc-Parking-107102928126130/

9 https://www.macclesfield-live.co.uk/police-shocker-officers-catch-speeding-15866814

10 https://www.cheshire-live.co.uk/news/chester-cheshire-news/bizarre-death-sold-here-graffiti-21447950

References

Ablitt, J. and James Smith, R. (2019). Working out Douglas's aphorism: Discarded objects, categorisation practices, and moral inquiries. *Sociological Review*, 67(4), 866–85.

Agnew, R. (2020). The ordinary acts that contribute to ecocide: A criminological analysis. In A. Brisman and N. South (eds), *Routledge international handbook of green criminology* (2nd edn, pp 52–67). Routledge.

Ajil, A., Jendly, M. and Campistol Mas, C. (2020). 'Yes, *security*, there is security. But other than that nothing': An empirical enquiry into the 'everyday (in)security' of Syrian and Iraqi refugees in Jordan. *British Journal of Criminology*, 60(6), 1395–415.

Amin, A. (2006). The good city. *Urban Studies*, 43(5–6), 1009–23.

Amin, A. and Thrift, N. (2016). *Seeing like a city*. Polity.

Amin, A. and Lancione, M. (2022). Introduction: Thinking cities from the ground. In A. Amin and M. Lancione (eds), *Grammars of the urban ground* (pp 1–26). Duke University Press.

Anand, N. (2017). *Hydraulic city: Water and the infrastructures of citizenship in Mumbai*. Duke University Press.

Anderson, B. (2022). Forms and scenes of attachment: A cultural geography of promises. *Dialogues in Human Geography*, 13(3), 392–409.

Angelo, H. (2019). The greening imaginary: Urbanized nature in Germany's Ruhr region. *Theory and Society*, 48, 645–69.

Anker, E. (2023). *Ugly freedoms*. Duke University Press.

Annavarapu, S. (2025). Surface tensions: Roads, potholes and the embodied politics of driving in urban India. *International Journal of Urban and Regional Research*. https://doi.org/10.1111/1468-2427.70017

Appleyard, B. (1980). Liveable streets: Protected neighborhoods? *The Annals of the American Academy of Political and Social Science*, 451, 106–17.

Arroyo-Quiroz, I. and Wyatt, T. (eds) (2018). *Green crime in Mexico: A collection of case studies*. Palgrave.

Back, L. (2015). Why everyday life matters: Class, community and making life livable. *Sociology*, 49(5), 820–36.

Bahceci, S., Bradford, B., Girling, E., Loader, I. and Sparks, R. (2023). Unsettled crossings: Underpass journeys in an English town. *Criminological Encounters*, 6(1), 81–94.

Balzacq, T., Léonard, S. and Ruzicka, J. (2016). 'Securitization' revisited: Theory and cases. *International Relations*, 30(4), 494–531.

Barker, A. and Holmes, G. (2021). *What makes a park feel safe or unsafe? The views of women, girls and professionals in West Yorkshire*. Available at: https://eprints.whiterose.ac.uk/id/eprint/194214/1/Parks%20Report%20FINAL%207.12.2022.pdf

Bateson, G. (1972). *Steps to an ecology of mind: Collected essays in anthropology, psychiatry, evolution, and epistemology*. University of Chicago Press.

Bawaka Country, Wright, S., Suchet-Pearson, S., Lloyd, K., Burarrwanga, L., Ganambarr, R., Ganambarr-Stubbs, M., Ganambarr, B., Maymuru, D., and Sweeney, J. (2016). Co-becoming Bawaka: Towards a relational understanding of place/space. *Progress in Human Geography*, 40(4), 455–75.

Benati, L. (2007). *The 'great moderation' in the United Kingdom*. Working Paper Series 769, European Central Bank. Available at: https://ideas.repec.org/p/ecb/ecbwps/2007769.html

Benson, E.S. (2020). *Surroundings: A history of environments and environmentalism*. University of Chicago Press.

Bentley, J. (2019). *Autopia: The future of cars*. Atlantic.

Berg, J. and Shearing, C. (2011). The practice of crime prevention: Design principles for more effective security governance. *South African Crime Quarterly*, 36, 23–30.

Berg, J. and Shearing, C. (2018). Governing-through-harm and public goods policing. *The Annals of the American Academy of Political and Social Science*, 679(1), 72–85.

Berg, J. and Shearing, C. (2024). 'Everything old is new again': Private urban security governance responses to new harmscapes. *Urban Studies*, 61(4), 2703–18.

Berlant, L. (2016). The commons: Infrastructures for troubling times. *Environment and Planning D: Society and Space*, 34(3), 393–419.

Bertolini, L. (2020). From 'streets for traffic' to 'streets for people': Can street experiments transform urban mobility? *Transport Reviews*, 40(6), 734–53.

Beveridge, R., Naumann, M. and Rudolph, D. (2024). The rise of 'infrastructural populism': Urban infrastructure and right-wing politics. *Geography Compass*, e12738. https://doi.org/10.1111/gec3.12738

Bhan, G., Keith, M., Parnell, S. and Pieterse, E. (2025). *Cities rethought: A new urban disposition*. Polity.

Bittner, E. (1990). *Aspects of police work*. Northeastern University Press.

Blair, T. (2005). *Speech to Labour Party conference*. Available at: https://www.theguardian.com/uk/2005/sep/27/labourconference.speeches

Blaustein, J., Miccelli, M., Hendy, R. and Hutton Burns, K. (2023). Resilience policing and disaster management during Australia's Black Summer bushfire crisis. *International Journal of Disaster Risk Reduction*, 95, 103848.

Blaustein, J., Shearing, C. and Miccelli, M. (2024). Adaptive policing for a climate crisis. *Policing and Society*, 1–13.

Bloch, S. (2021). Policing car space and the legal liminality of the automobile. *Progress in Human Geography*, 45(1), 136–55.

Bookchin, M. (1982). *The ecology of freedom*. Cheshire Books.

Bottoms, A. (2007). Place, space, crime, and disorder. In M. Maguire, R. Morgan and R. Reiner (eds), *The Oxford handbook of criminology* (4th edn, pp 528–74). Oxford University Press.

Bottoms, A. (2012). Developing socio-spatial criminology. In M. Maguire, R. Morgan and R. Reiner (eds), *The Oxford handbook of criminology* (5th edn, pp 450–89). Oxford University Press.

Bottoms, A. and Wiles, P. (2002). Environmental criminology. In M. Maguire, R. Morgan and R. Reiner (eds), *The Oxford handbook of criminology* (3rd edn, pp 620–56). Oxford University Press.

Bradford, B., Girling, E., Loader, I. and Sparks, R. (2024). Policing and sense of place: 'Shallow' and 'deep' security in an English town. *British Journal of Criminology*, 64(4), 791–810.

Bradford, B., Girling, E., Loader, I. and Sparks, R. (2025a). 'Seeing disorder' in an English town. *Criminology & Criminal Justice*. https://doi.org/10.1177/17488958251342733

Bradford, B., Rowlands, D., Weirich, C. and Crawford, A. (2025b). What service should the police provide? Towards a minimum policing standard. *Policing and Society*, 35(7), 1001–18.

Braga, A. and Bond, B. (2008). Policing crime and disorder hot spots: A randomized controlled trial. *Criminology*, 46(3), 577–607.

Brand, C. (2024). Confronting mobesity is vital for the global electrification of transport. *Nature Energy*, 9, 909.

Braun, R. and Randell, R. (2022). The vermin of the street: The politics of violence and the *nomos* of automobility. *Mobilities*, 17(1), 53–68.

Brewster, B. and Mayerfeld Bell, M. (2022). The environmental Goffman: Towards an environmental sociology of everyday life. *Society and Natural Resources*, 23, 45–57.

Brisman, A. (2004). Double whammy: Collateral consequences of conviction and imprisonment for sustainable communities and the environment. *Willian & Mary Environmental Law and Policy Review*, 28(2), 423–75.

Brisman, A. (2007). Crime-environment relationships and environmental justice. *Seattle Journal for Social Justice*, 6(2), Article 14.

Brisman, A. and South, S. (2014). *Green cultural criminology: Constructions of environmental harm, consumerism, and resistance to ecocide*. Routledge.

Brisman, A. and South, N. (2018). Green criminology and environmental crimes and harms. *Sociology Compass*, 13(1), e12650.

Brisman, A. and South, N. (eds) (2020). *The Routledge international handbook of green criminology* (2nd edn). Routledge.

Brisman, A., McClanahan, B., South, N. and Walters, R. (2018). *Water, crime and security in the twenty-first century: Too dirty, too little, too much*. Palgrave.

Burawoy, M. (2003). Revisits: An outline of a theory of reflexive ethnography. *American Sociological Review*, 68, 645–79.

Burua, M. (2023). *Lively cities: Reconfiguring urban ecology*. University of Minnesota Press.

Busby, J. (2022). *States and nature: The effects of climate change on security*. Cambridge University Press.

Canning, V. and Tombs, S. (2021). *From social harm to zemiology: A critical introduction*. Routledge.

Ceccato, V., Sundling, C. and Gliori, G. (2025). *Transit safety for a sustainable city: People, paths and places in railway stations*. Springer.

Chakraborty, D. (2021). *The climate of history in a planetary age*. University of Chicago Press.

Charbonnier, P. (2021). *Affluence and freedom: An environmental history of political ideas*. Polity.

Churchill, D., Crawford, A. and Barker, A. (2017). Thinking forward through the past: Urban order in (Victorian) public parks. *Theoretical Criminology*, 22(4), 523–44.

Clarke, R.V. (1995). Situational crime prevention. *Crime and Justice*, 19, 91–150.

Clarke, R.V. and Hough, M. (1984). *Crime and police effectiveness*. The Stationery Office.

Cohen, L.E. and Felson, M. (1979). Social change and crime rate trends: A routine activity approach. *American Sociological Review*, 44, 588–608.

Cozens, P.M. (2011). Urban planning and environmental criminology: Towards a new perspective for safer cities. *Planning Practice and Research*, 26(4), 481–508.

Cozens, P.M., Saville, G. and Hillier, D. (2005). Crime prevention through environmental design (CPTED): A review and modern bibliography. *Property Management*, 328–56.

Crawford, A. (2018). Temporalities in security: Long-term sustainability, the everyday and the emergent in the Anthropocene. In C. Holley and C. Shearing (eds), *Criminology and the Anthropocene* (pp 153–80). Routledge.

Crawford, A. and Hutchinson, S. (2016). Mapping the contours of 'everyday security': Time, space and emotion. *British Journal of Criminology*, 56(5), 1184–202.

Cresswell, T. (1996). *In place/out of place: Geography, ideology and transgression*. University of Minnesota Press.

Crow, G. (2018). *What are community studies?* Bloomsbury.

Crowe, T. (2000). *Crime prevention through environmental design: Applications of architectural design and space management concepts.* Butterworth Heinemann.

Crutzen, P. and Stoermer, E. (2000). The 'Anthropocene'. *Global Change Newsletter*, 41, 17–18.

Curtice, J., Scholes, A. and Chabdu, A. (2025). *Britain's democracy: A health check*. National Centre for Social Research.

Davies, P.A. (2014). Green crime and victimization: Tensions between social and environmental justice. *Theoretical Criminology*, 18(3), 300–16.

Davies, T. (2022). Slow violence and toxic geographies: 'Out of sight' to whom? *Environment and Planning C: Politics and Space*, 40(2), 409–27.

Di Ronco, A. and South, N. (2026). Green activism and social networks of care, repair and action. In C. Bald and M. Herrero (eds), *The Routledge international handbook of criminal justice social work* (pp 112–22). Routledge.

Dobson, J. (2017). From 'me towns' to 'we towns': Activist citizenship in UK town centres. *Citizenship Studies*, 21(8), 1015–33.

Doughty, K., Hu, H. and Smit, J. (2022). Therapeutic landscapes during the COVID-19 pandemic: Increased and intensified interactions with nature. *Social & Cultural Geography*, 24(3–4), 661–79.

Douglas, M. (1975). *Implicit meanings: Essays in anthropology*. Routledge.

Dryzek, J. and Pickering, J. (2019). *The politics of the Anthropocene.* Oxford University Press.

Eckersley, R. (2004). *The green state: Rethinking democracy and sovereignty*. MIT Press.

Elmstrøm Lauritsen, J. and Schultz, N. (2026). Geo-social classes and planetary existentialism: An interview with Nikolaj Schultz. *Distinktion: Journal of Social Theory*. https://doi.org/10.1080/1600910X.2026.2630666

Erker, S., Coyle, D. and Westwood, A. (2023). *Townscapes: A universal basic infrastructure for the UK*. Bennett Institute for Public Policy.

Fassin, D. (2013). Why ethnography matters: On anthropology and its publics. *Cultural Anthropology*, 28(4), 621–46.

Felski, R. (2000). *Doing time: Feminist theory and postmodern culture*. New York University Press.

Ferrell, J. (2013). Tangled up in green: Cultural criminology and green criminology. In N. South and A. Brisman (eds), *Routledge international handbook of green criminology* (pp 349–64). Routledge.

Fisher, L. (2017). *Environmental law: A very short introduction*. Oxford University Press.

Gandy, M. (2022). Urban political ecology: A critical reconfiguration. *Progress in Human Geography*, 46(1), 21–43.

Garland, D. (2001). *The culture of control: Crime and social order in contemporary society*. Oxford University Press.

Gerbaudo, P. (2021). *The great recoil: Politics after populism and pandemic*. Verso.

Giddens, A. (1991). *Modernity and self-identity*. Polity.

Girling, E., Loader, I. and Sparks, R. (2000). *Crime and social change in middle England: Questions of order in an English town*. Routledge.

Girling, E., Bradford, B., Loader, I. and Sparks, R. (2026). The greening of environmental criminology: Changing meanings of disorder in an English town. *British Journal of Criminology*, 66(3), 477–94.

Goldstein, D. (2010). Toward a critical anthropology of security. *Current Anthropology*, 51(4), 487–517.

Gorz, A. (1973). *The social ideology of the motor car*. Available at: https://atlasofplaces.com/essays/the-social-ideology-of-the-motorcar/

Goyes, D. (2019). *Southern green criminology: A science to end ecological discrimination*. Bristol University Press.

Goyes, D. and South, N. (2017). Green criminology before 'green criminology': Amnesia and absences. *Critical Criminology*, 25(2), 165–81.

Goyes, D., South, N., Astroina Abaibira, M., Baicué, P., Cuchimba, P. and Tatiana Ramos Ñeñetofe, D. (2021). Genocide and ecocide in four Colombian Indigenous communities: The erosion of a way of life and memory. *British Journal of Criminology*, 61(4), 965–84.

Graham, S. and Thrift, N. (2007). Out of order: Understanding repair and maintenance. *Theory, Culture & Society*, 24(3), 1–25.

Groombridge, N. (2013). Matter all over the place: Litter, criminology and criminal justice. In N. South and A. Brisman (eds), *The Routledge international handbook of green criminology* (1st edn, pp 394–408). Routledge.

Gupta, A. (2021). Infrastructure as decay and the decay of infrastructure. In G. Hage (ed), *Decay* (pp 37–46). Duke University Press.

Hage, G. (2021). Introduction: States of decay. In G. Hage (ed), *Decay* (pp 1–16). Duke University Press.

Hall, T. and Smith, R.J. (2015). Care and repair and the politics of kindness. *Sociology*, 49(1), 3–18.

Hamilton, J. (2013). *Security: Politics, humanity and the philology of care.* Princeton University Press.

Haraway, D. (2016). *Staying with the trouble: Making kin in the Chthulucene.* Duke University Press.

Harcourt, B. (2001). *Illusion of order: The false promise of broken windows policing.* Harvard University Press.

Harrington, J. and Shearing, C. (2017). *Security in the Anthropocene: Reflections on safety and care.* Transcript.

Harris, J. (2022). Disastrous cuts loom, but the Tories have a secret weapon – our weary fatalism. *The Guardian*, 30 October.

Hitchen, E. (2016). Living and feeling the austere. *New Formations*, 87, 102–18.

Holley, C. and Shearing, C. (eds) (2018). *Criminology and the Anthropocene.* Routledge.

Honig, B. (2017). *Public things: Democracy in disrepair.* Fordham University Press.

Hope, T. and Sparks, R. (2000). For a sociological theory of situations (or how useful is pragmatic criminology?). In A. von Hirsch, D. Garland and A. Wakefield (eds), *Situational crime prevention: Ethics and social context* (pp 175–92). Hart.

Hopkins, R. (2008). *The transition handbook: From oil dependency to local resilience.* Green Books.

Horne, R., Ambrose, L., Walker, G. and Nelson, A. (eds) (2024). *Post-carbon inclusion: Transitions built on justice.* Bristol University Press.

Huijbens, E. (2021). *Developing earthly attachments in the Anthropocene*. Routledge.

Intergovernmental Panel on Climate Change (2012). *Special report on managing risks of extreme events and disasters to advance climate change adaptation*. Cambridge University Press.

Jacobs, J. (1961). *The death and life of great American cities*. Vintage.

Jarvis, L. (2019). Towards a vernacular security studies: Origins, interlocutors, contributions, and challenges. *International Studies Review*, 21, 107–26.

Johnson, K., Wyatt, T., Coulthard, S. and O'Neill, C. (2020). Littering in the north-east of England: A sign of social disorganisation? In A. Brisman and N. South (eds), *The Routledge international handbook of green criminology* (2nd edn, pp 658–76). Routledge.

Jupp, E. (2022). *Care, crisis and activism: The politics of everyday life*. Policy Press.

Katz, C. (1996). Towards minor theory. *Environment and Planning D: Society and Space*, 14, 487–99.

Kelsey, T. and Kenny, M. (2021). *The value of social infrastructure*. Bennett Institute for Public Policy.

Kern, L. (2020). *Feminist city: Claiming space in a man-made world*. Verso.

Kirby, J. (2017). Fleshing out an ecological habitus. *Nature and Society*, 21(2), 89–114.

Klein, N. (2014). *This changes everything: Capitalism vs. the climate*. Penguin.

Klinenberg, E. (2018). *Palaces for the people: How social infrastructure can help fight inequality, polarization, and the decline of civic life*. Random House.

Kocisky, K. (2022). Towards conceptions of green gentrification as more-than-human. *Environment and Planning E: Nature and Space*, 5(2), 646–65.

Krause, M. (2021). *Model cases: On canonical research objects and sites*. University of Chicago Press.

Kurnicki, K. (2021). What do cars do when they are parked? Material objects and infrastructuring in social practices. *Mobilities*, 17(1), 37–52.

Lam, A., South, N. and Brisman, A. (2022). A convergence of crises: COVID-19, climate change and bunkerization. *Crime, Media, Culture*, 19(3), 327–44.

Larkin, B. (2013). The politics and poetics of infrastructure. *Annual Review of Anthropology*, 42, 327–43.

Latour, B. (2017). *Facing Gaia: Eight lectures on the new climatic regime*. Polity.

Latour, B. (2018). *Down to Earth: Politics in the new climatic regime*. Polity.

Lerner, J. (2014). *Urban acupuncture*. Island Press.

Lloyd-Rose, M. (2023). *Into the night: A year with the police*. Picador.

Loader, I. (2006). Policing, recognition and belonging. *The Annals of the American Academy of Political and Social Science*, 605(1), 201–21.

Loader, I. (2025). Concerning cars: Automobility and the contours of control, order and harm. *Annual Review of Criminology*, 8, 215–37.

Loader, I. (2026). Beyond motonormative punishment: On road safety as environmental regulation. *British Journal of Criminology*. https://doi.org/10.1093/bjc/azag026

Loader, I. and Walker, N. (2007). *Civilizing security*. Cambridge University Press.

Loader, I., Sparks, R., Bradford, B., Casey, R., Girling, E. and Polanska, G. (2023). Security and everyday life in uncertain times. In A. Liebling, L. McAra and S. Maruna (eds), *The Oxford handbook of criminology* (7th edn, pp 567–86). Oxford University Press.

Loader, I., Bradford, B., Girling, E., Sparks, R. and Bahceci, S. (2025). Inescapable objects? Automobility and everyday disorder in an English town. *American Journal of Cultural Sociology*. https://doi.org/10.1057/s41290-025-00256-w

Longstaff, A., Willer, J., Chapman, J., Czarnomski, S. and Graham, J. (2015). *Neighbourhood policing: Past, present and future*. Police Foundation.

Lucas, C. and Reid, R. (2025). It's time for climate populism. *New Statesman*, 7 February.

Lundberg, K. (2022). Walking at the edges of green criminology: The edges of the city and the extraordinary consequences of ordinary harms. *Criminological Encounters*, 5(1), 18–34.

Lundberg, K., Popovski, H. and Young, A. (2024). Justice in the streets? Towards a criminology of more-than-spatial justice. *Criminology & Criminal Justice*. https://doi.org/10.1177/1748895824125

Lynch, M. (1995). The greening of criminology: A perspective on the 1990s. *Critical Criminologist*, 2(3), 1–15.

Lynch, M. and Boggess, L.N. (2015). Ecocities, crime and justice: Ecocity theory, social disorganization, and green criminology. *Sociological Spectrum*, 35(4), 309–28.

Lynch, M. and Long, M. (2022). Green criminology: Capitalism, green crime and justice, and environmental destruction. *Annual Review of Criminology*, 5, 255–76.

Malik, A. and Berg, J. (2024). Polycentric governance, epistocracy and the limits of policing knowledge in preparing for the climate crisis. *Policing and Society*, 1–15.

Marquet, O., Anguelovski, I., Nello-Deakin, S. and Honey-Rosés, J. (2024). Decoding the 15-minute city debate: Conspiracies, backlash, and dissent in planning for proximity. *Journal of the American Planning Association*, 91(1), 117–25.

Marres, N. (2026). The postnormal society: On trials of co-existence in everyday environments. https://osf.io/preprints/socarxiv/xupgf_v1

Marsh, P., O Diekmann, L., Egerer, M., Lin, B., Ossola, A. and Kingsley, J. (2021). Where birds felt louder: The garden as a refuge during COVID-19. *Wellbeing, Space and Society*, 2, 100055.

Martin, C. (2024). Average weight of new cars rises by nearly 400 kg in seven years. *Autocar*. Available at: https://www.autocar.co.uk/car-news/new-cars/average-weight-new-cars-rises-nearly-400kg-seven-years

Massey, D. (1991). A global sense of place. *Marxism Today*, June, 24–29.

Massey, D.S. and Denton, N.A. (1993). *American apartheid: Segregation and the making of the underclass*. Harvard University Press.

Massey-Chase, B., Akehurst, S. and Alvis, S. (2026). *Apathy and opposition: Understanding the real threats to net zero*. Institute for Public Policy Research.

Matczak, A. (2025). *Adapting to climate change in modern policing: The rise of the 'eco-cop'*. Springer.

Matczak, A. and Bergh, S.I. (2023). A review of the (potential) implications of climate change for policing practice worldwide. *Policing*, 17, 1–8.

Matthews, D. (2023). *Earthbound: The aesthetics of sovereignty in the Anthropocene*. Edinburgh University Press.

McClanahan, B. and South, N. (2020). 'All knowledge begins with the senses': Towards a sensory criminology. *British Journal of Criminology*, 60(1), 3–23.

McDonald, M. (2021). *Ecological security*. Cambridge University Press.

Meek, J. (2025). Ten-foot chopsticks. *London Review of Books*, 4 December.

Miles, C., Broad, R. and Oldham, M. (2026). The safety of women who run: Negotiating fear, experiences and the normalisation of abuse. *British Journal of Criminology*, 66(1), 111–29.

Millie, A. (2022). Guerilla gardening as normalised law-breaking: Challenges to land ownership and aesthetic order. *Crime, Media, Culture*, 19(2), 191–208.

Mitchell, D. (2005). The S.U.V. model of citizenship: Floating bubbles, buffer zones and the rise of the 'purely atomic' individual. *Political Geography*, 24, 77–100.

Mouffe, C. (2022). *Towards a green democratic revolution: Left populism and the power of affects*. Verso.

Mould, O. (2023). From globalisation to the planetary: Towards a critical framework of planetary thinking in geography. *Geography Compass*, 17(9), e12720.

Murphy, A. (2012). 'Litterers': How objects of physical disorder are used to construct subjects of social disorder in a poor suburb. *The Annals of the American Academy of Political and Social Science*, 642(1), 210–27.

Mutongwizo, T., Holley, C., Shearing, C. and Simpson, N. (2019). Resilience policing: An emerging response to shifting harm landscapes and reshaping community policing. *Policing*, 15(1), 606–21.

Mutongwizo, T., Blaustein, J. and Shearing, C. (2021). Resilience policing and climate change: Adaptive responses to hydrological emergencies. *CrimRxiv*, 1–19.

Neocleous, M. (2008). *Critique of security*. Edinburgh University Press.

Newburn, T. and Ward, S. (2022). *Orderly Britain: How Britain has resolved everyday problems from dog fouling to double parking*. Robinson.

New Economics Foundation (2024). *Trapped behind the wheel: How England's new builds lock us into car-dependency*. Available at: https://neweconomics.org/2024/11/trapped-behind-the-wheel

Niven, A. (2023). *The north will rise again: In search of the future in northern heartlands*. Bloomsbury.

Norgaard, K.M. (2011). *Living in denial: Climate change, emotions and everyday life*. MIT Press.

Nyman, J. (2021). The everyday life of security: Capturing space, practice, and affect. *International Political Sociology*, 15(3), 313–37.

Office for National Statistics (2025). *Crime in England and Wales, year ending 2025*. Available at: https://www.ons.gov.uk/peoplepopulationandcommunity/crimeandjustice/bulletins/crimeinenglandandwales/yearendingmarch2025

Olsson, L., Jerneck, J., Thoren, H., Persson, J. and O'Byrne, D. (2015). Why resilience in unappealing to social science: Theoretical and empirical investigations of the scientific use of resilience. *Science Advances*, 1(4), e1400217. DOI:10.1126/sciadv.1400217

O'Reilly, K. (2012). Ethnographic returning, qualitative longitudinal research and the reflexive analysis of social practice. *The Sociological Review*, 60(3), 518–36.

Otto, F. (2025). *Climate injustice: Why we need to fight global inequality to combat climate change*. Greystone Books.

Paris, H. (2001). Human security: Paradigm shift or hot air? *International Security*, 26(2), 87–102.

Park, R. (1936). Human ecology. *American Journal of Sociology*, 42(1), 1–15.

Park, R. (1952). *Human communities: The city and human ecology*. Free Press

Paterson, M. (2007). *Automobile politics: Ecology and cultural political economy*. Cambridge University Press.

Pearsall, H. (2018). New directions in urban environmental/green gentrification research. In L. Lees and M. Phillips (eds), *Handbook of gentrification studies* (pp 329–45). Edward Elgar.

Pemberton, S. (2016). *Harmful societies: Understanding social harm*. Palgrave.

Persak, N. (2021). Offending sights and urban governance: Expectations of city aesthetics and spatial responses to the unsightly. In N. Persak and A. di Ronco (eds), *Harm and disorder in the urban space: Social control, sense and sensibility* (pp 52–77). Routledge.

Rajan, S. (1996). *The enigma of automobility: Democratic politics and pollution control*. University of Pittsburgh Press.

Read, R. (2022). *Why climate change matters*. Bloomsbury.

Róin, T., Petersen, M.S. and Róin, Á. (2025). 'I can still drive a car': Self-presentation in later life and the symbolic value of driving. *Ageing and Society*, 45(1), 164–75.

Rosa, H. (2019). *Resonance: A sociology of our relationship to the world*. Polity.

Sampson, R. (2012). *Great American city: Chicago and the enduring neighbourhood effect*. University of Chicago Press.

Sampson, R., Raudenbush, S.W. and Earls, F. (1997). Neighborhoods and violent crime: A multilevel study of collective efficacy. *Science*, 277(5328), 918–24.

Sayer, A. (2011). *Why things matter to people: Social science, values and ethical life*. Cambridge University Press.

Schoolman, E.D. (2020). Building community, benefiting neighbors: 'Buying local' by people who do not fit the mould for 'ethical consumers'. *Journal of Consumer Culture*, 20(3), 285–304.

Schuilenberg, M. (2015). *The securitization of society: Crime, risk and social order*. New York University Press.

Schwanen, T. (2021). Achieving just transitions to low-carbon urban mobility. *Nature Energy*, 6(7), 685–87.

Schwanen, T., Hopkins, D. and Loader, I. (2026). Mobility freedoms: Conceptions of freedom in contestation over urban transport. *Urban Studies*. https://journals.sagepub.com/doi/10.1177/00420980251403399

Schwartz, A. and Jax, K. (2011). Etymology and original sources of the term 'ecology'. In A. Schwartz and K. Jax (eds), *Ecology revisited: Reflecting on concepts, advancing science* (pp 145–8). Springer.

Sennett, R. (2018). *Building and dwelling: Ethics for the city*. Penguin.

Shaw, C.R. and McKay, H.D. (1942). *Juvenile delinquency and urban areas: A study of rates of delinquency in relation to differential characteristics of local communities in American cities*. University of Chicago Press.

Sheller, M. (2004). Automotive emotions: Feeling the car. *Theory, Culture & Society*, 21(4–5), 221–42.

Simon, J. (2007). *Governing through crime*. Oxford University Press.

Simone, A. and Pieterse, E. (2017). *New urban worlds: Inhabiting dissonant times*. Polity.

Skogan, W. (1992). *Disorder and decline: Crime and the spiral of decay in American neighborhoods*. University of California Press.

Small, M. and Calarco, M. (2022). *Qualitative literacy: A guide to evaluating ethnographic and interview research*. University of California Press.

Smith, A., Whitten, M. and Ernwein, M. (2025). De-municipalisation? Legacies of austerity for England's public parks. *The Geographical Journal*, 191(3), e12158.

Smith, J.R. and Hall, T. (2017). Everyday territories: Homelessness, outreach work and city space. *British Journal of Sociology*, 69(2), 372–90.

Smith, R.L. and Pimm, S.L. (2025). Ecology. *Encyclopaedia Britannica*. https://www.britannica.com/science/ecology

Sollund, R.A. (2015). Introduction: Critical green criminology – an agenda for change. In R.A. Sollund (ed), *Green harms and crimes: Critical criminology in a changing world* (pp 1–26). Palgrave.

Solnit, R. (2014). *The Encyclopaedia of trouble and spaciousness*. Trinity University Press.

South, N. (1998). A green field for criminology? A proposal for a perspective. *Theoretical Criminology*, 2(2), 211–33.

South, N. (2014). Green criminology: Reflections, connections, horizons. *International Journal for Crime, Justice and Social Democracy*, 3(2), 5–20.

Starke, P., Lund Elbek, L. and Wenzelburger, G. (eds) (2025). *Unequal security: Welfare, crime and social inequality*. Routledge.

Stears, M. (2021). *Out of the ordinary: How everyday life inspired a nation and how it can again*. Belknap.

Steele, W., Hillier, J., MacCallum, D., Byrne, J. and Houston, D. (2021). *Quiet activism: Climate action at the local scale*. Macmillan.

Steffen, W., Broadgate, W., Deutsch, L., Gaffney, O. and Ludwig, C. (2015). The trajectory of the Anthropocene: The great acceleration. *The Anthropocene Review*, 2(1), 81–98.

Stengers, I. (2015). *In catastrophic times: Resisting the coming barbarism*. Open Humanities Press.

Stewart, M. and Eborall, H. (2025). 'The mortar between the bricks of the services': How third sector staff's risk work supported people who were homeless to access healthcare during the COVID-19 pandemic. *Health, Risk & Society*, 1–17.

Taub, R., Garth Taylor, D. and Dunham, J. (1984). *Paths of neighborhood change: Race and crime in urban America*. University of Chicago Press.

Taylor, R.B. and Covington, J. (1988). Neighborhood changes in ecology and violence. *Criminology*, 26, 553–89.

Taylor, R.B., Haberman, C.P. and Groff, R. (2019). Urban park crime: Neighborhood context and park features. *Journal of Criminal Justice*, 64, 101622.

Thaler, M. (2022). *No other planet: Utopian visions for a climate-changed world*. Cambridge University Press.

Tibbets, S.G. and Hemmens, C. (2018). *Criminological theory: A text/reader* (3rd edn). SAGE.

Tiratelli, M. (2025). Attitudes to crime and punishment in England and Wales, 1964–2023: A reinterpretation of the 1980s and a model of interactions between concern, punitiveness and prioritization. *The British Journal of Criminology*, 65(2), 387–404.

Traill, H., Anderson, S., Shaw, D., Cumbers, A. and McMaster, R. (2024). Caring at the edges: Infrastructures of care and repair in urban deprivation. *Environment and Planning D: Society and Space*, 42(3), 190–210.

Tronto, J. (1993). *Moral boundaries: A political argument for an ethic of care*. Routledge.

Tronto, J. (2013). *Caring democracy: Markets, equality and justice*. New York University Press.

Tsing, A. (2015). *The mushroom at the end of the world: On the possibility of life in capitalist ruins*. Princeton University Press.

Turner, V. (1970). *The forest of symbols: Aspects of Ndembu ritual*. Cornell University Press.

United Nations Development Programme (1994). *Human development report 1994: New dimensions of human security*. United Nations.

Valverde, M. (2011). Questions of security: A framework for research. *Theoretical Criminology*, 15(1), 3–22.

Valverde, M. (2014). Studying the governance of crime and security: Space, time, and jurisdiction. *Criminology & Criminal Justice*, 14(4), 379–39.

Walker, B., Holling, C., Carpenter, S. and Kinzig, A. (2004). Resilience, adaptability and transformability in socio-ecological systems. *Ecology and Society*, 9(2), 5.

Walker, I., Tapp, A. and Davis, A. (2023). Motonormativity: How social norms hide a major public health hazard. *International Journal of Environment and Health*, 11(1), 21–33.

Walker, J.T. (ed) (2011). *Social, ecological and environmental theories of crime*. Routledge.

Wallace-Wells, D. (2019). *The uninhabitable Earth: A story of the future*. Penguin.

Walters, R. (2014). Air pollution and invisible violence. In P. Davies, P. Francis and T. Wyatt (eds), *Invisible crimes and social harms: Critical criminological perspectives* (pp 142–69). Palgrave.

Weisburd, D., Bernasco, W. and Bruinsma, G. (eds) (2009). *Putting crime in its place: Units of analysis in geographic criminology*. Springer.

Weisburd, D., Hinkle, J.C., Famega, C. and Ready, J. (2011). The possible 'backfire' effects of hot spots policing: An experimental assessment of impacts on legitimacy, fear and collective efficacy. *Journal of Criminal Law and Criminology*, 7, 297–320.

Wenger, E. (1998). *Communities of practice: Learning, meaning and identity*. Cambridge University Press.

Whatmore, S. (2002). *Hybrid geographies: Natures, cultures, spaces*. SAGE.

White, R. (2008). *Crimes against nature: Environmental criminology and ecological justice*. Routledge.

White, R. (2014). *Environmental harm: An eco-justice perspective*. Bristol University Press.

White, R. (2018). Carbon criminals, ecocide and climate justice. In C. Holley and C. Shearing (eds), *Criminology and the Anthropocene* (pp 50–80). Routledge.

White, R. (2020). *Climate change criminology*. Bristol University Press.

White, R. (2022). *Theorising green criminology: Selected essays*. Routledge.

Wikström, P.-O. (2009). Social ecology of crime. In B. Huebner (ed), *Oxford bibliographies in criminology*. Oxford Academic. https://doi.org/10.1093/obo/9780195396607-0027

Williams, M. (2016). Care-full justice in the city. *Antipode*, 49(3), 821–39.

Williams, R. (2006). *The politics of modernism: Against the new conformists*. Verso.

Wilson, J.Q. and Kelling, G. (1982). Broken windows: The police and neighbourhood safety. *The Atlantic*, March.

Wilson, W.J. (1987). *The truly disadvantaged: The inner city, the underclass, and public policy*. University of Chicago Press.

Wo, J.C. and Rogers, E.M. (2024). Urban greenspace and neighborhood crime. *Criminology*, 62, 236–275.

Yarwood, R., Hall, T. and Ceccato, C. (2025). Southern criminology and place: Perspectives from human geography. In R.P. Cavalcanti, D. Fonseca, V. Vegh Weiss, K. Carrington, R. Hogg and J. Scott (eds), *The Palgrave handbook of criminology and the global south*. Palgrave. https://doi.org/10.1007/978-3-031-74932-2_18-1

Zerubavel, E. (2006). *The elephant in the room: Silence and denial in everyday life*. Oxford University Press.

Zerubavel, E. (2018). *Taken for granted: The remarkable power of the unremarkable*. Princeton University Press.

Index

References to tables appear in **bold** type; references to endnotes show both the page number and the note number (129n4). As Macclesfield is the main topic of the book, only general topics have been indexed under this heading: readers are advised to search using specific topics (for example, 'drug-related issues').

D

E

F